# THE ORIGINS OF THE CONFLICT BETWEEN SCIENCE AND RELIGION IN ANCIENT GREECE

Craig A. Leisy

# THE ORIGINS OF THE CONFLICT BETWEEN SCIENCE AND RELIGION IN ANCIENT GREECE


ISBN Number: 978-1-60571-132-4

Printed in the United States of America

Cover Credit: The Acropolis in Athens. Hand-painted with glazes by Vickie Leisy

# Idle Thoughts

Rain's only wet
if you're in it;
Yesterday's not the same
when you visit.

Chickadees don't have knees
and they gossip a lot;
pine trees just stand around
and complain they're too hot.

You can chat with dogs and cats,
they know a little English;
cats have trouble with vowels
but clearly pronounce meows.

I don't know why the housefly
is one of god's creations,
Darwinists insist, the fly
is one of man's relations.

You can prove the Rule of Three[1]
with principles of geometry,
and it's easy, when the moon is full,
to neatly square a circle.[2]

You can read every page
or skip to the end,
but tomorrow's just a place
when you get there.

---

[1] Refers to Lewis Carroll's *The Hunting of the Snark* and not the law of ratios.

*Just the place for a Snark! I have said it twice:*
*That alone should encourage the crew.*
*Just the place for a Snark! I have said it thrice:*
*What I tell you three times is true.*

The law of ratios allows the computation of an unknown if the other three values are given: If $\frac{A}{B} = \frac{C}{D}$ then, $A = \frac{BC}{D}$

[2] Squaring a circle refers to a method for computing a more accurate value for pi ($\pi$) which is the ratio of the circumference of a circle to its diameter: $\pi$ = circumference/diameter = 3.14.

# Dedication

For my son Corey. From an early age, he had a fascination with Greek mythology and it helped reawaken my own interest in ancient Greek history and philosophy. We discussed many of the ideas in this book as I worked on it.

# PREFACE

This monograph was originally intended to serve as an introduction to a planned larger work on the conflict between science and religion. It is written for the general reader who has limited familiarity with the subject. Liberal footnotes are provided for the reader who wishes to learn more about some aspect of this short history. Ancient sources are quoted extensively throughout so that they may, "speak in their own words."

Greek astronomy, an early science, introduced a new way of thinking and ultimately upset the foundations of popular religion. There are, of course, parallels in our present age that make this subject all the more relevant.

This is not meant to be a comprehensive history of the conflict between science and religion in ancient Greece. Rather, the outline presented here merely introduces the topic. Science is the search for natural causes to explain phenomena and these inquiries comprised a new philosophy that was distinct from, and in conflict with, religion.

I apologize for any errors of fact or composition. Every effort was made to carefully edit the manuscript.

# CONTENTS


Preface i

Contents ii

Introduction 1

The conflict between science and religion had its origins in ancient Greece about 500 BCE shortly after the first sciences appeared.

Chapter 1 The Origins of Science 5

Science was born in Miletus, a Greek colony in Ionia, then spread to mainland Greece and especially Athens.

Chapter 2 The History of Early Greek Astronomy

A brief survey of the principal discoveries of astronomy and their influence on popular religion in ancient Greece.

Thales of Miletus 12
Anaximander of Miletus 15
Anaximenes of Miletus 18
Pythagoras of Samos 18
Anaxagoras of Clazomenae 20
Empedocles of Acragas 23
Philolaus of Croton 24
Leucippus of Miletus 24
Democritus of Abdera 25
Eudoxus of Cnidos 25
Heraclides of Pontus 26
Aristarchus of Samos 27
Archimedes of Syracuse 30
Eratosthenes of Cyrene 31
Hipparchus of Nicea 36
Poseidonius of Apamea 39

## Chapter 3 The Beginnings of Religious Unbelief

A short history of the sources of religious unbelief and their impacts on society in ancient Greece.


| | |
|---|---|
| Xenophanes of Colophon | 45 |
| Anaxagoras of Clazomenae | 47 |
| Hippo of Rhegium | 52 |
| Heraclitus of Ephesus | 53 |
| Diogenes of Sinope | 54 |
| Diagoras of Melos | 56 |
| Protagoras of Abdera | 58 |
| Democritus of Abdera | 60 |
| Critias of Athens | 62 |
| Prodicus of Ceos | 64 |
| Bion of Borysthenes | 66 |
| Euripides of Salamis | 67 |
| Socrates of Athens | 78 |
| Aristotle of Stagira | 83 |
| Euhemerus of Messene | 85 |
| Epicurus of Samos | 88 |
| Other Religious Unbelief | 92 |


## Chapter 4 Sources of Conflict with Religion

A summary of other sources of conflict with religion in ancient Greece.


| | |
|---|---|
| Greek Religion | 94 |
| The State and Religion | 95 |
| City Life and Religion | 98 |
| Trade and Religion | 99 |
| History and Religion | 100 |
| Philosophy and Religion | 102 |
| The Sophists and Religion | 104 |
| Literature and Religion | 106 |
| Moral Criticism of Religion | 107 |

The Problem of Evil 110
Fate and Religion 117
Religious Unbelief 119
Science and Religion 124

Chapter 5 Conclusion 131

Remarks about the influence of the conflict between science and religion in ancient Greece.

Bibliography 137

Appendices

Early Greek Astronomy 143
Early Greek Unbelief 148

Index 153

About the Author 159

# INTRODUCTION

The conflict between science and religion in modern times had its origin in ancient Greek astronomy. The discovery that the heavenly bodies were subject to natural laws challenged the early myths about the gods which are the foundation of the western religions.

The modern conflict between science and religion actually began in ancient Greece. Religion is as old as man but science was born in Greece just 2,500 years ago. Science originated with the discovery of causal relationships that helped to explain occurrences in the natural world such as the predictable transit of the heavenly bodies across the skies, regular seasonal changes in weather, natural disasters, and many other events of major importance in the daily life of people. Some of the first sciences were astronomy, botany, zoology, biology, medicine, and geology – but astronomy is usually considered the oldest science.

Astronomy was of singular importance because the regular movement of the stars led to the development of calendars that were used to determine when to plant and harvest crops. This was vital knowledge in the agrarian economy of ancient Greece and even earlier civilizations in Babylonia and Egypt. For instance, in *Works and Days*, the Greek poet Hesiod (c. 700 BCE)[3] advises his brother Perses when to harvest grapes and when to squeeze them to make wine.

---

[3] Explanation of terms: c. – circa (approximate date), f. – flourished (usually mid-life), BCE – Before the Common Era (formerly B.C.), CE – Common Era (formerly A.D.).

But when Orion and Sirius are come into mid-heaven, and rosy-fingered Dawn sees Arcturus, then cut off all the grape clusters, Perses, and bring them home. Show them to the sun ten days and ten nights: then cover them over for five, and on the sixth day draw off into vessels the gifts of joyful Dionysus. But when the Pleides and Hyades and strong Orion begin to set, then remember to plough in season: and so the completed year will pass beneath the earth. [4]

Some scholars question whether Greece was the birthplace of astronomy. There is no doubt that the Greek astronomers learned from Egypt and Babylonia.[5] For example, the Babylonians observed the motions of the fixed stars and had extensive records of lunar and solar eclipses. The Babylonians used their astronomical observations to develop the practice of astrology in order to try to predict future events caused by the gods.

---

[4] Hesiod. *Works and Days*, lines 609-617 in *Hesiod: The Homeric Hymns and Homerica* (trans. by Hugh G. Evelyn-White) London: William Heinemann Ltd. (1982), p. 49. Hesiod offers advice to his brother Perses. The Pleides, or Seven Sisters (daughters of Atlas), rise in May and set in October. Ancient sailors preferred trips by sea during these months when the weather was moderate. See, M. Cary and E. H. Warmington. *The Ancient Explorers* Baltimore, MD: Penguin Books Rev. ed. (1963) [orig. published 1929], p.21

"The most forbidding feature of the Mediterranean is the frequency of its winter gales, which change direction rapidly and create incalculable cross-seas in confined waters. In ancient times these gales effectively closed the seas for all ordinary sailings from October to April, so that the season for navigation was confined to summer."

[5] The Jewish historian Josephus credits the Biblical figure Abraham (from Chaldea) with introducing astronomy to the Egyptians who, in turn, taught the Greeks. See, H. St. J. Thackeray (trans.). *Josephus* Vol. IV "Jewish Antiquities", Books 1-IV London: William Heinemann Ltd., Book I, 167-168 (p. 83):

"Thus gaining [the Egyptians] admiration…as a man of extreme sagacity, gifted not only with high intelligence but with power to convince his hearers on any subject which he undertook to teach, he introduced them to arithmetic and transmitted to them the laws of astronomy. For before the coming of Abraham the Egyptians were ignorant of these sciences, which thus traveled from the Chaldaeans into Egypt, whence they passed to the Greeks."

> ...the influence which the stars exerted upon our world seemed undeniable. Did not the rising and setting of the sun every day bring heat and cold, as well as light and darkness? Did not the changes of the seasons correspond to a certain state of the sky? What wonder, therefore, that by induction men arrived at the conclusion that even the lesser stars and their conjunctions had a certain connection with the phenomena of nature and the events of human life...Everything in sky and earth alike is incessantly changing, and it was thought that there existed a correspondence between the movements of the gods above and the alterations which occurred here below. This is the fundamental idea of astrology.[6]

The Egyptians developed accurate solar calendars. However, most of what we recognize as the foundation of modern astronomy was discovered by the Greeks. This includes: the plurality of worlds, the sphericity of the earth, the cause of lunar and solar eclipses, the rotation of the earth on its axis, measurement of the circumference of the earth, calculation of the distances to the sun and the moon, and development of the heliocentric hypothesis. [7] This monograph will briefly trace the

---

[6] Franz Cumont. *Astrology and Religion Among the Greeks and Romans* New York: G. P. Putnam's Sons (1912), pp. 17-18.

[7] Copernicus is usually acknowledged as the author of the heliocentric hypothesis but Aristarchus of Samos asserted that the earth orbited the sun in the third century BCE as documented in the writings of Archimedes. The original manuscript of *De Revolutionibus* by Copernicus contained the following statement that was ultimately deleted from the first printed edition:

"And if we should admit that the course of the sun and moon could be demonstrated even if the earth is fixed, then with respect to the other wandering bodies there is less agreement. It is credible that, for these and similar causes (and not because of the reason of motion, which Aristotle mentions and rejects), Philolaus was aware of the mobility of the earth, and some even say that Aristarchus of Samos was of the same opinion." In Owen Gingerich. "From Copernicus to Kepler: Heliocentrism as Model and as Reality" *Proceedings of the American Philosophical Society* Vol. 117, No. 6 Symposium on Copernicus (December 31, 1973), p. 513.

development of ancient Greek astronomy from Thales (6th c. BCE) to Poseidonius (1st c. BCE) and assess its influence on the religious beliefs of those times.

# CHAPTER 1
# THE ORIGINS OF SCIENCE

It is marvelous to what length the depravity of man's intellect will go…And when they have dared to guess the distances of the sun from the earth they apply the same figures to the sky, on the ground that the sun is at its centre, with the consequence that they have at their finger's ends the dimensions of the world also. For they argue that the circumference of a circle is 22/7 times its diameter, as though the measure of the heavens were merely regulated from a plumbline![8]

## MILETUS

The first great Greek philosophers, whose investigations and writings constitute the origins of modern science, were residents of Miletus. Miletus was one of twelve Greek colonies in Ionia which were located on the Mediterranean coast of the modern state of Turkey. [9] Thales, Anaximander, Anaximenes, and Leucippus flourished in Miletus during pre-Socratic times beginning in the sixth century before the common era. Scholars have long speculated on the reasons why Miletus was the birthplace of modern science. There is general agreement that Miletus was a busy trading center because it was located on the sea trade routes of the eastern Mediterranean.

…Thales, Anaximander and Anaximenes were citizens of Miletus, an Ionian Greek city on the west coast of Asia Minor, from the beginning of the sixth century onwards. In their time Miletus, which had already existed for some five hundred years, was a center radiating tremendous energy. Ancient tradition hailed it as the

---

[8] Gaius Plinius Secundus. *Natural History* (trans. By H. Rackham) 10 vols. Cambridge, MA: Harvard University Press (1967), Loeb Classical Library, Book II, 87, p. 231.

[9] The city-states included: Chios, Samos, Phocaea, Colophon, Clazomenae, Ephesus, Erythrae, Priene, Teos, Myus, Lebedos, and Miletus.

mother of no less than ninety colonies, and modern research confirms the reality of about forty-five of them – in itself an astonishing number. One of the oldest of these was the commercial settlement of Naucratis in Egypt, founded in the middle of the seventh century. Miletus possessed great wealth, which it obtained both by acting as a trading center for materials and manufactured goods brought to the coast from inner Anatolia and by the export of a variety of manufactures of its own. Milesian woolen goods were famous throughout Greek lands thus shipping, trade, and industry combined to give this busy harbor-city a leading position and wide connections, extending to the Black Sea in the north, Mesopotamia in the east, Egypt in the south, and the Greek cities of south Italy in the west. Its government was aristocratic, and its leading citizens lived in an atmosphere of luxury and a culture which may be broadly described as humanistic and materialistic in tendency. Its high standard of living was too obviously the product of human energy, resource and initiative for it to acknowledge any great debt to the gods.[10]

The popular historian Will Durant observed that Miletus, during the sixth century BCE, was the wealthiest Greek city.

It was in this stimulating environment that Greece first developed two of its most characteristic gifts to the world – science and philosophy. The crossroads of trade are the meeting place of ideas, the attrition ground of rival customs and beliefs; diversities beget conflict, comparison, thought; superstitions cancel one another, and reason begins. Here in Miletus as later in Athens, were men from a hundred scattered states; mentally active through competitive commerce, and freed from the bondage of tradition by long absences from their native altars and homes. Milesians themselves traveled to distant cities, and had their eyes opened by civilizations of Lydia, Babylonia, Phoenicia, and Egypt; in this way, among others, Egyptian geometry and Babylonian astronomy entered the Greek mind. Trade and mathematics, foreign commerce and geography, navigation and astronomy, developed hand in hand.[11]

---

[10] W.K.C Guthrie. *A History of Greek Philosophy* Cambridge, England: Cambridge University Press (1962) Vol. I "The Earlier Presocratics and the Pythagoreans", pp. 29-30.

[11] Will Durant. *The Story of Civilization: Part II The Life of Greece* New York: Simon and Schuster (1939), p. 134.

According to the ancient geographer Strabo of Amasia (63 BCE-21 CE), Miletus probably began as a Cretan colony and derived its name from a city of the same name on the northeastern coast of the island of Crete. [12] Miletus established its own colonies on the shores of the Black Sea and in Egypt beginning in the eighth century BCE. [13]

> Miletus was said to have been the mother of ninety colonies in the different seas. It is difficult to believe that she had the manpower to people so many cities single-handed. And we may rather think of her as the leader in a broader movement of overseas expansion. Miletus herself will have provided 'oecists' (founders), shipping, and the requisite knowledge of sea-routes and local conditions; but other Ionic cities must have provided a good proportion of the emigrants. [14]

Miletus was mainly a trading center connecting ports all along the eastern Mediterranean but it also exported its own products consisting of olive oil, figs, fish, woolen garments, pottery, purple dyes, and wooden furniture.

> ...the so-called triad of wheat, olives and wine, was central to the ancient agriculture, economy and society of Miletos as it was in other contemporary Greek poleis (city-states). The three main crops of the

---

[12] Horace Leonard Jones (trans.). *The Geography of Strabo* Vol.VI London: William Heinemann Ltd. (1960), p. 205 [Loeb Classical Library].

[13] George Sarton. *A History of Science: Ancient Science Through the Golden Age of Greece* Cambridge, MA: Harvard University Press (1952), p. 183. According to Sarton, "During [the rule of Ahmose II in Egypt (569-525 BCE)], the Greek merchants had obtained so much power that they were allowed to build, or to reorganize, the city of Naucratis, on the Canopic branch of the Nile, not very far west of the capital, Sais. That city became the main center of Greek commerce in Egypt (somewhat like Alexandria in later Ptolemaic days)."

[14] J. M. Cook. *The Greeks in Ionia and the East* New York: Frederick A. Praeger (1963), p. 51. Pliny,in his *Natural History*, commented that Miletus established 90 colonies [N.H., 5.122].

triad are harvested at different times of the agricultural calendar, making them an ideal combination.[15]

The historian Adelaide Dunham asserted that the prosperity of Miletus was likely due to sea borne commerce and did not derive from the caravan trade routes connecting East and West. [16] However, recent archeological finds in excavations at Miletus suggest that one route of the "Silk Road" may have terminated on the coastline at Miletus. [17]

Historian John Burnet attributed the birth of science in Ionia primarily to the fact that colonists left behind everything – even their gods – and started anew. [18] This opinion found support from Martin Nilsson, a historian of Greek religion, who believed that the anthropomorphic religion promoted by the poems of Homer actually originated in Ionia.

---

[15] Alan M. Greaves. *Miletos: A History* London: Routledge (2002), pp. 15-16. See also p. 25:
"Olive oil was an essential of ancient Greek life, yet this ubiquitous commodity was rarely mentioned in literary sources. Not only was it used for cooking and eating but it also fueled lamps, formed the base for perfumes, preserved food and was used in place of soap. It may also have been used in the processing of textiles and wool and the byproducts of pressing could also be used as fuel, animal feed or fertilizer".

[16] Adelaide Glynn Dunham. *The History of Miletus: Down to the Anabasis of Alexander* London: University of London Press (1915), p. 11. "The theory has been held that Miletus was one of the chief outlets for the caravan trade with the East, and that to this circumstance she owed a great part of her wealth; but a review of the various overland trade routes proves this suggestion to be unsatisfactory."

[17] Alan M. Greaves. *Miletos: A History* London: Routledge (2002), pp. 10-11.

[18] John Burnet. *Early Greek Philosophy* 4th ed. Cleveland, OH: The World Publishing Co. (1964), p. 14 [reprint of original published in 1892].

The Homeric poems were created among the emigrants in the colonies on the coast of Asia Minor. The emigrants could not take with them the graves of their ancestors, consequently they could not continue the grave cult. It ceased of necessity, and thence followed a weakening of this important part of the religion. Its decline was furthered by the lack of introspection and the absence of tradition associated with life in the colonies, where every man was the shaper of his own fortunes; the tendency of the Greek temperament towards self-assertion and rationalism had free course. [19]

This theme, that the first Greek settlers in the Ionian colonies established new religious traditions that reflected the circumstances of their struggle to start new lives, found support with other historians. For instance, F. M. Cornford concluded that the Ionian Greeks were moved to substitute materialism for the natural religions of the homes that they left.

Nilsson insisted that the poems of Homer, by ascribing human characters and weaknesses to the gods eventually caused many to abandon the popular religion.

Homer's anthropomorphism gave rise to the first criticism of religion, and for the development of the Greek mind it had an importance the full extent of which has never been realized. For this humanizing of the gods served to ward off the conception of divine power as the magical, wonder-working agency which prevails in many religions, for instance, the Egyptian. Under this all-compelling magical power of the gods man bows in fear and terror, but from its fetters the Homeric humanization of the gods delivered the Greeks. They could henceforth of their own accord and by their own efforts find order and coherence in the world. From this origin came Greek science. [20]

---

[19] Martin P. Nilsson. *A History of Greek Religion* (trans. By F. J. Fielden) 2nd ed. New York: W. W. Norton & Co. (1964), p. 136 [originally published in 1925]

[20] Ibid., p. 179. See also, J. M. Cook. *The Greeks in Ionia and the East* New York: Frederick A. Praeger (1963), pp. 41-42.

"Whether in their arguments on Olympus or in their dealings with human beings, the gods of the *Iliad* are curiously lacking in better feelings, and lacking even in moral principle. They cannot often be held to set a noble example to mankind by their conduct, and almost

The theory that the development of science was largely made possible by the existence of a weak system of religious beliefs is supported by Burnet who concluded that the growth of superstition and the establishment of a priestly class may have prevented science from originating in Egypt or Babylonia.[21] The Egyptians replaced lunar calendars with more accurate solar calendars for their agriculture while the Babylonians used astronomical knowledge for astrology. There is no evidence that celestial observations were used for navigation, mapping or other geographical purposes until the Greeks began naming and cataloging the stars. The Greek development of astronomy eventually challenged the cosmogony of popular religion. According to Burnet,

> The Babylonians recorded celestial phenomena for astrological purposes, not for any scientific interest. There is no evidence that they attempted to account for what they saw in any but the crudest way. The Greeks on the other hand, made at least three discoveries of capital importance in the course of two or three generations.In the first place, they discovered that the earth is a sphere and does not rest on anything. In the second place, they discovered the true theory of lunar and solar eclipses; and in close connexion with that, they came to see, in the third place, that the earth is not the center of our system but revolves round the center like the planets. Not much later, certain

---

every one of them is at one time or another made to appear ridiculous. The divine scenes in the *Iliad* are generally introduced either as amusing interludes or as a machinery to precipitate the action; and the gods are reflections of human nature and human relationships. It is difficult to believe –though many scholars do somehow believe it – that Homer's brilliant, cynical treatment of his gods can really have concealed any deep religious conviction. His belief was in man, not in gods. In his universe, faith – in the sense of religious faith – is lacking; but there is hope in this world, and charity abounds.

[21] John Burnet. *Early Greek Philosophy* 4th ed. Cleveland, OH: The World Publishing Co. (1964), pp.3-4.

Greeks took, at least tentatively, the final step of identifying the center round which the earth and planets revolve with the sun. [22]

To be sure, the Babylonians were the first to make a careful record of the movements of the heavenly bodies and the occurrence of celestial phenomena such as eclipses. However, it was the Greeks who tried to make sense of it.

Babylonian astronomers had long recognized the periods of the planets on the basis of records kept over many centuries, and they were able to predict their exact movements on the basis of number series. The Greeks took the data from them and also translated the names of the gods which the Babylonians had used to designate the planets as Hermes, Aphrodite, Ares, Zeus, and Kronos – in the corresponding Latin forms they are current even today, Mercury, Venus, Mars, Jupiter, and Saturn; but the geometric formulation with the intuitive spatial model and the conclusions drawn from it are the work of the Greeks. [23]

Miletus was the birthplace of other 'firsts' besides the sciences. Hecataios of Miletus (fl. c. 500 BCE) was considered the 'father of geography'. He wrote books on descriptive geography. His maps showed the Mediterranean Sea surrounded by the three land masses: Europe, Asia and Africa. [24] Cadmos of Miletus (fl. c. 540 BCE), who authored histories of Ionia, is sometimes considered the 'earliest Greek historian.' [25]

---

[22] Ibid., p.23

[23] Walter Burkert. *Greek Religion: Archaic and Classical* (trans. by John Raffan) Oxford: Basil Blackwell Ltd. (1985), p. 326.

[24] D. R. Dicks (ed). *The Geographical Fragments of Hipparchus* London: University of London (1960), p. 20.

[25] George Sarton. *A History of Science: Ancient Science through the Golden Age of Greece* Cambridge, MA: Harvard University Press (1952), pp. 186, 193.

# CHAPTER 2
# THE HISTORY OF EARLY GREEK ASTRONOMY

O mighty heroes, of loftier than mortal estate, who have discovered the law of those great divinities [the sun and stars] and released the miserable mind of man from fear...all hail to your genius, ye that interpret the heavens and grasp the facts of nature, discoverers of a theory whereby you have vanquished gods and men![26]

Astronomy developed in the sixth century before the common era in Ionian Greece. To be sure, the Greeks learned of the cycle of eclipses from the Babylonians and adopted accurate solar calendars devised by the Egyptians. However, it was the Greeks that produced a cosmology to explain the nature and origins of the universe and invented mathematical astronomy using their knowledge of geometry.

## THALES OF MILETUS

Thales of Miletus (624-547 BCE) was the first philosopher to make a discovery important to the development of ancient Greek astronomy. The historian Herodotus recorded that Thales predicted the year of a solar eclipse that occurred during a battle between the Lydians and the Medes on May 28, 584 BCE. According to Herodotus,

(W)ar broke out between the Lydians and the Medes and lasted for five years, and during this period sometimes the Medes won and sometimes the Lydians; there was also one night battle. As the war was proving to be a draw between the two peoples, in the sixth year it happened that during a fight, when the combatants were already

[26] Gaius Plinius Secundus (Pliny the Elder). *Natural History* (trans. By H. Rackham) 10 vols. Cambridge, MA: Harvard University Press (1967) [Loeb Classical Library], Book II, IX 53-55, pp. 203-205.

closely engaged, suddenly day became night. The occurrence of this eclipse of daylight had already been predicted to the Ionians by Thales of Miletus, and he had set as his limiting date the year in which the eclipse actually took place. But the Lydians and the Medes, when they saw night instead of day before their eyes, gave over the fight, and both were more eager on their own behalf to make peace.[27]

Observations by the Babylonians indicated that a solar eclipse occurs every 223 lunar cycles or 18 years and 11 days. This is known as the Saros period. It is possible that Thales learned of it from the Babylonians but most modern historians disagree with this explanation.[28]

---

[27] Herodotus. *The History* (trans. by David Grene) Chicago: The University of Chicago Press (1987), 1.74. According to Pliny the Elder, Gaius Plinius Secundus (23-79 CE), "The original discovery was made in Greece by Thales of Miletus, who in the fourth year of the 48th Olympiad (585 BCE) foretold the eclipse of the sun that occurred in the reign of Alyattes, in the 170th year after the foundation of Rome." See Pliny. *Natural History* 10 vols. (trans. by H. Rackham) Cambridge, MA: Harvard University Press [reprinted 1967] Loeb Classical Library, Book II, viii52-ix54. (p. 203).

[28] See, for example, George Sarton. *A History of Science: Ancient Science through the Golden Age of Greece* Cambridge, MA: Harvard University Press (1952), p. 171:

"That is a pretty legend, but it has become impossible to give credence to it. The theory was that the old Babylonians had discovered the saros, a period that enabled them to predict eclipses. Thales had heard of that in Egypt and might even have witnessed the Egyptian eclipse of 603 or been told of it. A new eclipse would then occur, or at least might occur, 223 synodic months or 18 years 11 days later, that is in 585…(I)t is now agreed by historians of ancient astronomy that the Babylonians could not have discovered that period before the fifth or fourth century. Hence, Thales could not have learned it from them."

Also, see John Burnet. *Early Greek Philosophy* 4th ed. London: Adam & Charles Black (1945), p.43

"We do not occupy a position at the centre of the earth, and the geocentric parallax has to be taken into account. It would only, therefore, be possible to tell by means of the (Saros) cycle that an eclipse of the sun would be visible somewhere, and that it might be worthwhile to look out for it, though an observer at a given place

Because Thales was an almost mythic figure, numerous 'firsts' were attributed to him by early Greek historians. Most of them are probably exaggerations or achievements of later philosophers. For example, Thales was credited by both Greek and Roman writers with discovery that the Earth was a sphere and calculating the size of the sun and moon.

> [Thales] was the first to determine the sun's course from solstice to solstice, and according to some the first to declare the size of the sun to be one seven hundred and twentieth part of the solar circle, and the size of the moon to be the same fraction of the lunar circle. [29]

An accomplishment by Thales in the development of primitive mathematics related to using the principle of similar triangles to compute the height of Egyptian pyramids.

---

might be disappointed five times out of six. Now if we may judge from reports by Chaldean astronomers which have been preserved, this was just the position of the Babylonians in the eighth century B.C. They watched for eclipses at the proper dates; and, if they did not occur, they announced the fact as a good omen."
And see, Patricia F. O'Grady. *Thales of Miletus* Burlington, VT: Ashgate Publishing Co. (2002), p. 131
"The period of 223 lunations equates to 6585.321124 days. This extra 0.321124 of a day means that each recurring solar eclipse will be visible approximately 115.6° to the west of the previous eclipse – that is just under one-third of the circumference of the Earth, being a period of time of almost 7.5 hours. There is no possibility that this regression to the west by almost one-third of the circumference of the Earth, and the time of 7.5 hours, could have been known to the ancient astrologers or astronomers, a fact which seems not to have been taken into account by the philosophers who attribute to Thales the prediction of the eclipse of 585 B.C. on the basis of the Saros. Nor for that matter, was it taken into account by those like Heath, who believe that, by the sixth century B.C., the Babylonians were able to predict solar eclipses through knowledge of that cycle".

[29] Diogenes Laertius. *Lives of Eminent Philosophers* [trans. by R. D. Hicks] 2 vols. London: William Heinemann (1925) Book I, Chapter 1, 24.

[Thales] had no instructors, except that he went to Egypt and spent some time with the priests there. Hieronymus informs us that he measured the height of the pyramids by the shadow they cast, taking the observation at the hour when our shadow is the same length as ourselves. [30]

Thales was considered by the ancient Greeks to be one of the Seven Sages or wisest men.

## ANAXIMANDER OF MILETUS

Anaximander of Miletus (611-546 BCE), a student or contemporary of Thales, is considered by some scholars to be the 'father of Greek natural philosophy' (i.e., science) because he used observations and reason to explain a broad range of phenomena in nature. [31] Previous to Anaximander, the forces in nature were attributed to the gods. His were some of the earliest Greek writing but none of his writings survive except as fragments quoted in books by later philosophers. His

---

[30] Ibid., Book I, Chapter 1, 27. See also, Note for pp. 335, 336 from Additional Notes (p. xxxii) in H. F. Tozer. *A History of Ancient Geography* 2nd ed. London: Cambridge University Press (1935):
"The philosopher Thales (c. 600 B.C.) was said to have ascertained the height of an Egyptian pyramid by measuring (1) the shadow of the pyramid from the half-way point on its base, (2) his own shadow, (3) his own height. A simple application of the rule-of-three gave him the height of the pyramid.
[NOTE: The Rule of Three is a ratio of proportions where a/b = c/d and therefore a = bc/d].

[31] See, for instance, Charles H. Kahn. *Anaximander and the Origins of Greek Cosmology* New York: Columbia University Press (1960), p. 7:
What the system of Anaximander represents for us is nothing less than the advent, in the West at any rate, of a rational outlook on the natural world. This new point of view asserted itself with the total force of a volcanic eruption, and the ensuing flood of speculation soon spread from Miletus across the length and breadth of the lands in which Greek was spoken.

ideas influenced the Milesian school in Ionia then quickly spread to the Greek mainland.

Anaximander believed that the earth had a cylindrical shape with a length three times its depth. He reasoned that the earth did not need to be supported by some medium because it was the same distance from all points in the universe and therefore would not fall like other bodies.

> …the earth is a heavenly body supported by nothing but remaining in its place by reason of its equal distance from everything.[32]

The ramifications of this novel idea for religion were truly revolutionary.

> The consequences which the new cosmology brought about for religious thought were profound and extensive…In the new scheme of the world, which put the earth hanging free in space, there was no room for the underworld, the realm of the dead under the surface of the earth.[33]

---

[32] Hippolytus. *Philosophumena or the Refutation of All Heresies* Book I, 7 (trans. by F. Legge) London: Society for Promoting Christian Knowledge (1921). See also, Dirk L. Couprie. "The Discovery of Space: Anaximander's Astronomy" in Dirk L. Couprie, Robert Hahn and Gerard Naddaf. *Anaximander in Context* Albany, NY: State University of New York Press (2003), p. 207:
"Now we have to return to the question of how Anaximander possibly could have conceived of this idea. For…it was not something he could have observed. Anaximander must have somehow concluded that the earth hangs unsupported in space…When Anaximander concluded from the daily appearance and disappearance of the celestial bodies that they make full circles and go underneath the earth as well, he must have drawn his consequence from that, namely, that the earth must be unsupported in the center of those circles."
[33] Gilbert Murray. *Four Stages of Greek Religion* New York: Columbia University Press (1912), p. 101.

Anaximander is usually credited with creating the first map of the inhabited areas of the earth that is thought to have consisted of a circular land mass surrounded by seas and rivers. [34]

> [Anaximander] was the first to draw on a map the outline of land and sea, and he constructed a globe as well. [35]

It is generally accepted that the Milesians borrowed heavily from Babylonian celestial observations and particularly records of past lunar and solar eclipses (see Thales).

> In astronomy the Babylonians might amass data extending over centuries, based on careful observation and involving considerable ingenuity of calculation. But it did not occur to them to use this mass of data as the basis for constructing a rational cosmology like that of Anaximander or Plato. This gift for abstraction, with its limitless possibilities and (we must add) its inherent danger, was the peculiar property of the Greeks.[36]

The Babylonians also invented the use of the gnomon or sun dial. Anaximander was said to have used the gnomon to determine the seasons. Later Greek astronomers depended on the gnomon to determine the latitude of cities for more accurate maps. A gnomon was even used to calculate the circumference of the earth.

Anaximander believed in a boundless or infinite universe with a plurality of worlds similar to the earth.

---

[34] For an excellent concise summary of the contributions of early Greek astronomers, see Thomas L. Heath. *Greek Astronomy* New York: Dover Publications Inc. (1991) [reprint of 1932 edition].

[35] Diogenes Laertius. *Lives of Eminent Pholosophers* [trans. by R. D. Hicks] 2 vols. London: William Heinemann (1925) Book II, Chapter 1, 1.

[36] W.K.C. Guthrie. *A History of Greek Philosophy* Vol. I "The Earlier Presocratics and the Pythagoreans" Cambridge University Press (1962), p. 37.

These ideas persisted through all the early Greek astronomers. This was a break with the past where the heavenly bodies were considered gods.

## ANAXIMENES OF MILETUS

Anaximenes of Miletus (585-528 BCE) was a younger contemporary of Anaximander. Some credit him with using the gnomon to tell time.[37] Anaximenes stated that there were earth-like bodies that revolved around the earth with the stars. These invisible bodies helped explain why there were more lunar eclipses than solar eclipses. Anaximenes also said that the stars don't give off heat like the sun because of their great distance from the earth.[38]

Most other philosophers that made significant contributions to ancient astronomy lived in other Greek colonies in Asia Minor and Africa. Eventually, the Pythagorean school in Italy replaced the Milesian as the new source of cosmology in pre-Socratic times. There were several notable thinkers in the Pythagorean school including Parmenides and Philolaus.

## PYTHAGORAS OF SAMOS

Pythagoras of Samos (580-500 BCE) or, perhaps, Parmenides of Elea (504-450 BCE), is said to have

---

[37] According to Pliny, Anaximenes discovered the use of the gnomon for telling time. See -
Pliny. *Natural History* ( trans. by H. Rackham) Vol. I London: William Heinemann Ltd. (1938), Book II, v. 187 (p. 319): "This theory of shadows and the science of gnomonics was discovered by Anaximenes of Miletus, the pupil of Anaximander…he first exhibited at Sparta the time-piece they call 'Hunt-the-Shadow'."
[38] Hippolytus. Op. cit., Book I, 6.

discovered the sphericity of the earth.[39] The rounded shadow of the earth cast on the moon during a lunar eclipse was observed by the Greeks. Aristotle later commented on this in his book *On the Heavens*,

> ...the shapes which the moon itself each month shows are of every kind – straight, gibbous, and concave – but in eclipses, the outline is always curved: and, since it is the interposition of the earth that makes the eclipse, the form of this line will be caused by the form of the earth's surface, which is therefore spherical. [40]

In addition, Aristotle stated that the fact that different constellations were visible above the horizon at different geographical locations was further evidence of a curved surface. Aristotle also credited the Pythagoreans with determining that the earth was not located in the center of the universe.

> As to [the earth's] position, there is some difference of opinion. Most people - all, in fact, who regard the whole heaven as finite - say it lies at the centre. But the Italian philosophers known as Pythagoreans take the contrary view. At the centre, they say, is fire, and the earth is one of the stars, creating night and day by its circular motion about the centre.[41]

Many of the early astronomers attempted to estimate the distance from the earth to the moon and sun – often using principles of geometry as their method. These estimates were wildly inaccurate because observations

---

[39] Diogenes Laertius. *Lives of Eminent Philosophers* [trans. by R. D. Hicks] 2 vols. London: William Heinemann (1925) Book VIII, Chapter 1, 48-49.
"Further, we are told that [Pythagoras] was the first to call the heaven the universe and the earth spherical, though Theophrastus says it was Parmenides, and Zeno that it was Hesiod".

[40] Aristotle. *De Caelo* (On the Heavens) Book II, Chapter 14, 297b 25-30 (trans. by J. L. Stocks)
in Richard McKeon (ed). *The Basic Works of Aristotle* New York: Random House (1941).

[41] Ibid., Book II, Chapter 13, 293a.

and measurements depended upon primitive equipment. Pythagoras published his own estimates.

> The penetrating genius of Pythagoras ...inferred that the distance of the moon from the earth was 15,750 miles, and that of the sun from the moon twice that figure, and of the sun from the twelve signs of the zodiac three times.[42]

Pythagoras asserted that Venus was both the morning star and evening star. Previously, it was thought that they were different stars.

> Below the sun revolves a very large star named Venus, which varies its course alternately, and whose alternative names in themselves indicate its rivalry with the sun and moon – when in advance and rising before dawn it receives the name of Lucifer, as being another sun and bringing the dawn,whereas when it shines after sunset it is named Vesper, as prolonging the daylight, or as being a deputy for the moon. This property of Venus was first discovered by Pythagoras of Samos about the 42 Olympiad, 142 years after the foundation of Rome.[43]

## ANAXAGORAS OF CLAZOMENAE

Anaxagoras of Clazomenae (500-428 BCE) is recognized for several important discoveries. He believed that the moon was illuminated by the sun and that heavenly bodies were not gods but rather,

> ...the sun and moon and all the stars are burning stones...[44]

---

[42] Pliny. *Natural History* (trans. by H. Rackham) London: William Heinemann Ltd. (1967), p. 227.
[43] Ibid., pp. 191,193. The footnotes give the dates 612-609 BCE for the 42nd Olympiad and 753 BCE for the foundation of Rome.
[44] Hippolytus. Op. cit., Book I, 7
Also, Pliny, the famous Roman encyclopedist of the first century BCE, wrote an account of a meteorite striking the Earth near Aegospotami, in what is now the modern state of Turkey. According to him, Anaxagoras was said to have foretold the event. See Gaius Plinius Secundus (Pliny the Elder). *Natural History* (trans. By H,

Anaxagoras was the first to determine that the actual cause of a lunar eclipse was that the earth blocked the light of the sun and cast a shadow over the moon. He also taught that the moon's surface was earth-like with plains, ravines and mountains.[45] Like Anaximenes before him, Anaxagoras believed that there were other worlds and that they were inhabited. According to Plutarch, Anaxagoras was persecuted for the irreligious nature of his teachings.[46]

The first man to put in writing the clearest and boldest of all doctrines about the changing phases of the moon was Anaxagoras. But he was

---

Rackham) 10 vols. Cambridge, MA: Harvard University Press (1967) [Loeb Classical Library], Book II, LIX 149-150 (pp. 285-287):

"The Greeks tell the story that Anaxagoras of Clazomenae in the 2nd year of the 78th Olympiad was enabled by his knowledge of astronomical literature to prophecy that in a certain number of days a rock would fall from the sun; and that this occurred in the daytime in the Goat's River district of Thrace (the stone is still shown – it is of the size of a wagon-load and brown in colour), a comet also blazing in the nights at the time.

[45] Diogenes Laertius. *Lives of Eminent Philosophers* [trans. by R. D. Hicks] 2 vols. London: William Heinemann (1925) Book II, Chapter 3, 8.

[Anaxagoras] declared the sun to be a mass of red hot metal and to be larger than the Peloponnesus…he declared that there were dwellings on the moon, and moreover hills and ravines.

[46] There is an interesting anecdote by Plutarch regarding Anaxagoras in prison: "There is no place that can take away the happiness of a man, nor yet his virtue or wisdom. Anaxagoras, indeed, wrote on the squaring of the circle while in prison." From Plutarch *On Exile* 17,607E, F quoted in Ivor Thomas (trans.) *Selections Illustrating the History of Greek Mathematics* 2 vols. London: William Heinemann Ltd. (1980) [reprint of orig. published in 1939], Vol. I "From Thales to Euclid", p. 309. Anaxagoras, like many other early Greek mathematicians, attempted to solve the problem of "squaring the circle" or finding a square (or polygon) with the same area as a circle in order to simplify geometric calculations. At this time, nothing was known about trigonometry and angular measures were not yet in use.

no ancient authority, nor was his doctrine in high repute. It was still under seal of secrecy, and made its way slowly among a few only, who received it with a certain caution rather than with implicit confidence. Men could not abide the natural philosophers and "visionaries", as they were then called, for that they reduced the divine agency down to irrational causes, blind forces, and necessary incidents. Anaxagoras was with difficulty rescued from imprisonment by Pericles, and Socrates, though he had nothing whatever to do with such matters, nevertheless lost his life because of philosophy. [47]

## Diogenes Laertius, a compiler writing in the third century of the common era, described the trial of Anaxagoras on charges of impiety.

Of the trial of Anaxagoras different accounts are given. Sotion in his *Succession of the Philosophers* says that he was indicted by Cleon on a charge of impiety, because he declared the sun to be a mass of red-hot metal; that his pupil Pericles defended him, and he was fined five talents and banished. Satyrus in his *Lives* says that the prosecutor was Thucydides, the opponent of Pericles, and the charge one of treasonable correspondence with Persia as well as impiety; and that sentence of death was passed on Anaxagoras by default…Hermippus in his *Lives* says that [Anaxagoras] was confined in the prison pending his execution; that Pericles came forward and asked the people whether they had any fault to find with him in his own public career, to which they replied that they had not. "Well", he continued,

---

[47] Bernadotte Perrin (trans.). *Plutarch's Lives* 11 vols. London: William Heinemann Ltd. (1932) [Loeb Classical Library] Vol. III *Nicias*, Chapter 23, 2-3. One historian explains that Anaxagoras, "seems to have been the first philosopher to take up his abode at Athens; there he enjoyed the friendship of Pericles. When Pericles became unpopular shortly before the outbreak of the Peloponnesian War, he was attacked through his friends. Anaxagoras was accused of impiety for saying that the sun was a red-hot stone and the moon earth. Some say that he was fined and banished, others that it was intended to put him to death and that he was with difficulty saved by Pericles. Ultimately he went and lived at Lampsacus, where he died at the age of seventy-two." Thomas L. Heath *Greek Astronomy* New York: Dover Publications (1991)[reprint of 1932 edition], p. xxxiii.

"I am a pupil of Anaxagoras; do not then be carried away by slanders and put him to death. Let me prevail upon you to release him." [48]

Anaxagoras quit Athens for Lampsacus where he died. Diogenes Laertius reports that Anaxagoras regarded astronomy as the purpose of his life.

Being asked to what end he had been born, [Anaxagoras] replied, "To study sun and moon and heavens." [49]

## EMPEDOCLES OF ACRAGAS

The philosopher Empedocles (490-444 BCE) is principally remembered for two ideas. He conceived that the universe was comprised of combinations of four elements: earth, fire, water and air. This belief became widely held in Presocratic Greece. Another original idea attributed to Empedocles was that light requires time to move from the sun to the earth. Aristotle reported that:

---

[48] Diogenes Laertius. *Lives of Eminent Philosophers* [trans. by R. D. Hicks] 2 vols. London: William Heinemann (1925) Book II, Chapter 3, 12-13.
According to Josephus, Abraham believed that the heavenly bodies were not visible gods and, as a result, he was forced to flee Mesopotamia by the Chaldeans:
"[Abraham] was thus the first boldly to declare that God, the creator of the universe, is one, and that, if any other being contributed aught to man's welfare, each did so by His command and not in virtue of its own inherent power. This he inferred from the changes to which land and sea are subject, from the course of sun and moon, and from all the celestial phenomena; for, he argued, were these bodies endowed with power, they would have provided for their own regularity…It was in fact owing to these opinions that the Chaldaeans and the other peoples of Mesopotamia rose against him, and he, thinking fit to emigrate, at the will and with the aid of God, settled in the land of Canaan." H. St. J. Thackeray (trans.) *Josephus* London: William Heinemann Ltd. (1978) Vol. IV "Jewish Antiquities", Book I, 154-157 (pp.77-79).

[49] Ibid., 10.

Empedocles, for instance, says that the light from the sun reaches the intervening space before it reaches the eye on the earth. And this might well seem to be the fact. For, when a thing is moved, it is moved from one place to another, and hence a certain time must elapse during which it is being moved from one place to the other. But every period is divisible. Therefore there was a time when the ray was not seen, but was being transmitted through the medium. [50]

## PHILOLAUS OF CROTON

Another Pythagorean, Philolaus of Croton (470-390 BCE) believed that the earth, planets, sun and moon all revolved around a central fire (not the sun). [51]

At the center, (the Pythagoreans) say, is fire, and the earth is one of the stars, creating night and day by its circular motion about the centre. [52]

The Pythagoreans, like Anaximenes, believed that invisible dark bodies, even a counter-earth, orbited the central fire and accounted for the fact that lunar eclipses were more frequent than solar eclipses.

## LEUCIPPUS OF MILETUS

Leucippus of Miletus (fl. 450-420 BCE) lived about the same time as Anaxagoras and Empedocles. He is noteworthy for developing the 'atomist' cosmogony.[53]

---

[50] Aristotle. *De Sensu* , Ch. 6, 446a 26-b2.
[51] Diogenes Laertius. *Lives of Eminent Philosophers* [trans. by R. D. Hicks] 2 vols. London: William Heinemann (1925) Book VIII, Chapter 7, 85.
"[Philolaus] was the first to declare that the earth moves in a circle, though some say that it was Hicetas of Syracuse".
[52] Ibid., Book II, Chapter 13 293a 20-25.
[53] According to Strabo, "And if one must believe Poseidonius, the ancient dogma about atoms originated with Mochus, a Sidonian, born before the Trojan times. Strabo. *The Geography of Strabo* 8 vols. (trans. by Horace Leonard Jones) London: William Heinemann Ltd.

Leucippus believed that the infinite was a void that contained innumerable worlds formed by the collision of small particles of matter called atoms.

Leucippus is also credited with introducing the concept of causation which is the basis of all scientific investigation.

> To Leucippus we owe the sentence in which the universal rule of causation was proclaimed in unequivocal language: "Nothing happens without a cause, but everything with a cause and by necessity."[54]

## DEMOCRITUS OF ABDERA

Another famous atomist, Democritus of Abdera (c. 460-370 BCE), stated that the order of the heavenly bodies from earth was the moon, the sun, other planets, then the fixed stars. Like Leucippus, Democritus taught that there were an infinite number of worlds. However, he said that the worlds varied in size, some may have more than one sun or moon, and some are devoid of life.

> He said also that the ordered worlds are boundless and differ in size, and that in some there is neither sun nor moon, but that in others both are greater than with us, and yet others more in number. And that the intervals between the ordered worlds are unequal...And that some ordered worlds are bare of animals and plants and of all water. [55]

## EUDOXUS OF CNIDOS

Eudoxus of Cnidos (408 BCE-355 BCE), a student of Plato, postulated a system of rotating concentric spheres to explain the irregular motion of the planets. He

---

(1960) Vol. VII, 16.2.24 (p. 271).

[54] Theodor Gomperz. *Greek Thinkers: A History of Ancient Philosophy* 2 vols. (trans. By Laurie Magnus) London: John Murray (1920) Vol. I, p. 317 [orig. publ. 1896].

[55] Hippolytus. *Philosophumena*, Op. Cit., Book I, 11.

believed that each planet had four spheres about different axes causing a combined motion that appeared irregular in comparison with the revolution of the fixed stars. According to Eudoxus, the movement of the sun and moon can also be explained by spheres.

Eudoxus supposed that the motion of the sun or of the moon involves, in either case, three spheres, of which the first is the sphere of the fixed stars, and the second moves in the circle which runs along the middle of the zodiac, and the third in the circle which is inclined across the breadth of the zodiac; but the circle in which the moon moves is inclined at a greater angle than that in which the sun moves. [56]

Eudoxus was the first Greek astronomer to develop a geometric model to represent the motion of the planets.[57]

HERACLIDES OF PONTUS

Heraclides of Pontus (388 BCE-315 BCE), a student of Plato like Eudoxus, discovered that the earth rotated on its axis with one revolution per day. This rotation caused apparent motion of the sun across the skies

---

[56] Aristotle. *Metaphysica* (Metaphysics) Book XII, Chapter 8, 1073B 15-25 (trans. by W.D. Ross) in Richard McKeon (ed.) *The Basic Works of Aristotle* New York: Random House (1941)

[57] Giorgio de Santillana. *The Origins of Scientific Thought: 600 BC to 300 AD* Chicago,IL: University of Chicago Press (1961), p. 245. "The system was and remained a provisional working model. Eudoxus could not but know that Mars and Venus, at least, show startling variations in brightness...He must have known, too, the change in apparent diameter of the sun and moon through their cycle, which was noted very early. All this implied variations in distance from the earth which made it very awkward to speak of homocentric spheres."

during the day and similar motion of the planets and fixed stars during hours of darkness. Previous to Heraclides, it was assumed that the heavenly bodies moved while the earth was at rest. Simplicius, a commentator for Aristotle, stated,

> Heraclides of Pontus thought that he could save the phenomena by assuming that the earth is at the centre (of the universe) and moves in a circle, while the sky is at rest. [58]

Heraclides made a second important discovery. He observed that Mercury (Lucifer) and Venus (the Morning Star) both orbit the sun. Until this discovery, it was thought that all heavenly bodies orbited the earth or a central fire (see the Pythagoreans).

## ARISTARCHUS OF SAMOS

Aristarchus of Samos (310 BCE-230 BCE), a contemporary of the famous mathematician Archimedes, made perhaps the most important discovery of ancient Greek astronomy. He was the author of the 'heliocentric hypothesis' which proposed that the earth orbits the sun. The writings of Aristarchus, explaining the sun-centered system, have been lost. However, in *The Sand-Reckoner* by Archimedes, there is a brief reference to the radical hypothesis of Aristarchus.

> Now you are aware that universe is the name given by most astronomers to the sphere whose centre is the centre of the earth and whose radius is equal to the straight line between the centre of the sun and the centre of the earth. This is the common account, as you have heard from astronomers. But Aristarchus of Samos brought out a book consisting of some hypotheses, in which the premises lead to the result that the universe is many times greater than that now so

---

[58] Simplicius on Aristotle *De Caelo* (On the Heavens), Book II, Chapter 13, 293b 30. Quoted by H. B. Gottschalk *Heraclides of Pontus* Oxford: Clarendon Press (1980), p. 59.

called. His hypotheses are that the fixed stars and the sun remain unmoved, that the earthrevolves about the sun in the circumference of a circle, the sun lying in the middle of the orbit, and that the sphere of the fixed stars, situated about the same centre as the sun, is so great that the circle in which he supposes the earth to revolve bears such a proportion to the distance of the fixed stars as the centre of the sphere bears to its surface.[59]

There are other references to the heliocentric hypothesis of Aristarchus in the writings of Plutarch and Sextus Empiricus. The heliocentric hypothesis was rejected in favor of Ptolemy's geocentric (earth-centered) hypothesis by nearly all astronomers until Copernicus. [60]

There is an account by Plutarch that Aristarchus was charged with impiety for proposing that the earth orbited the sun:

...Cleanthes thought that the Greeks ought to lay an action for impiety against Aristarchus the Samian on the ground that he was disturbing the hearth of the universe because he sought to save the phenomena by assuming that the heaven is at rest while the earth is

---

[59] T. L. Heath (ed.) *The Works of Archimedes* New York: Dover Publications, Inc. "The Sand-Reckoner", pp. 221-222 [reprint of 1897 edition] Also, quoted by Ivor Thomas (trans.) *Selections Illustrating the History of Greek Mathematics: Vol. II Aristarchus to Pappus* Rev. London: Harvard University Press (1993) [The Loeb Classical Library].

[60] D. R. Dicks. *The Geographical Fragments of Hipparchus* London: University of London (1960), p.115. According to Dicks, Seleucus the Chaldean (Babylonian), c. 2nd century BCE, was the last ancient astronomer that was known to have accepted the heliocentric hypothesis of Aristarchus. The historian W. W. Tarn commented on the failure of Seleucus to prove the heliocentric theory of Aristarchus as follows:

"The pity of it was that, could heliocentricism have been established, it would have killed astrology and saved the world infinite trouble". W. W. Tarn. *Hellenistic Civilization* 2nd ed. London: Edward Arnold & Co. (1930), p. 263. The authority of Hipparchus and Ptolemy firmly established the geocentric theory until the time of Copernicus.

revolving along the ecliptic and at the same time is rotating about its own axis. [61]

Another significant work by Aristarchus, *On the Sizes and Distances of the Sun and Moon*, has survived. In it, Aristarchus uses geometry to calculate the relative sizes of the sun and moon and their distances from earth. By measuring the angle between the sun and moon from earth when the moon is exactly half-illuminated by the

---

[61] Harold Cherniss and William C. Helmbold (trans.) *Plutarch's Moralia* Vol. XII "Concerning the Face Which Appears in the Orb of the Moon", 923 (p.55). There is a note that, "'Against Aristarchus' appears in the list of Cleanthes' writings given by Diogenes Laertius, vii, 174." Also, see Plutarch "On the Face in the Moon", 6, 922F-923A trans. by Ivor Thomas in *Selections Illustrating the History of Greek Mathematics: Vol. II Aristarchus to Pappus* Rev. London: Harvard University Press (1993) [The Loeb Classical Library], p. 5. According to the historian Santillana, Copernicus, who was erroneously credited with authoring the heliocentric (sun-centered) hypothesis, first learned about the concept by reading Plutarch's *On the Face of the Moon*. "The dialogue *On the Face in the Round of the Moon* has had the extraordinary honor of inspiring both Copernicus and Newton. It is here that Copernicus found the mention of Aristarchus's system that he acknowledges in his book. It is here that Newton (who had read the dialogue in his adolescence) found an intimation of a gravitational force between the earth and the moon, which caused him to write later that there must have been an esoteric knowledge of universal attraction among the ancients, as evidenced in Plutarch." Giorgio de Santillana *The Origins of Scientific Thought: 600 BC to 300 AD*, Chicago, IL: University of Chicago Press (1961), p. 260. See also, Owen Gingerich. "From Copernicus to Kepler: Heliocentrism as Model and as Reality" *Proceedings of the American Philosophical Society* Vol. 117, No. 6, Dec. 31, 1973, p. 513. Gingerich quotes Copernicus from an early manuscript of *De revolutionibus* that was removed prior to printing: "And if we should admit that the course of the sun and moon could be demonstrated even if the earth is fixed, then with respect to the other wandering bodies there is less agreement. It is credible that, for these and similar causes (and not because of the reason of motion, which Aristotle mentions and rejects), Philolaus was aware of the mobility of the earth, and some even say that Aristarchus of Samos was of the same opinion."

sun, all angles of the triangle connecting the three bodies may be determined. [62] Aristarchus measured the angle formed by the earth, moon and sun as 87 degrees. [63] He calculated that the ratio of the distance of the sun from the earth to the distance of the moon from the earth was between 18:1 and 20:1. Aristarchus used this ratio and the apparent size of the earth's shadow during a lunar eclipse to calculate the ratios of the diameters of the sun and moon to that of the earth. He concluded that the ratios of the diameter of the sun and moon to the diameter of the earth were approximately 7:1 and 1/3:1 respectively. [64]

## ARCHIMEDES OF SYRACUSE

Archimedes of Syracuse (c. 287 BCE – 212 BCE) was a famous mathematician and astronomer. His work *The Sand-Reckoner*[65] includes a description of how to

---

[62] This assumes a right angle between the sun and earth from the moon.

[63] The angle is really 89 degrees 50 minutes. The small error in angular measure yields a huge error in the ratio of the distances. Aristarchus calculated the ratio of the distances of the sun from the earth to the distance of the moon from the earth as between 18:1 and 20:1. The actual value is 390:1 according to Albert Van Helden *Measuring the Universe: Cosmic Dimensions from Aristarchus to Halley* Chicago: The University of Chicago Press (1985), pp. 6-7.

[64] Ibid. p. 8. The actual ratio of the diameter of the sun to the earth is about 109:1 and the ratio of the diameter of the moon to the diameter of the earth is approximately 0.3:1. Average distances: earth-sun – 92,960,000 miles, earth-moon – 238,900 miles. Diameters: sun – 864,000 miles, moon – 2,160 miles; earth – 7,920 miles.

[65] T. L. Heath (ed.) *The Works of Archimedes* New York: Dover Publications, Inc. "The Sand-Reckoner" [reprint of 1897 edition]., pp. 221-232. Also, in Thomas L. Heath *Greek Astronomy* New York: Dover Publications, Inc. (1991), p. 109, "In the end Archimedes finds, by sheer calculation, that the number of grains of sand that would be contained in a sphere of the the size attributed to the universe is less than the number which we should express as $10^{63}$."

measure the diameter of the sun. His largest contribution to the development of mathematical astronomy was the calculation of a very accurate value for pi ($\pi$).[66] He concluded, by direct measurement, that pi was less than 1/164th part and greater than 1/200th part of a right angle.[67]

## ERATOSTHENES OF CYRENE

Eratosthenes of Cyrene (276 BCE- 195 BCE), a contemporary of Archimedes of Syracuse, was born in north Africa, educated in Athens, and eventually appointed royal librarian at Alexandria. He is best remembered as a geographer. Historian W. W. Tarn, citing Strabo, commented that Eratosthenes believed the presence of tides in the Mediterranean Sea proved that the oceans were all connected.

> The question of whether India and Africa joined or not, which had so perplexed Aristotle, had been settled by Alexander, and Eratosthenes' strongly critical mind never doubted that all the oceans were one and that the inhabited world, Europe-Asia-Africa, was an island; he pointed to the similarity of the tides in the Indian and Atlantic Oceans, and rightly deduced that one could sail from Spain round

---

Also, quoted by Ivor Thomas (trans.) *Selections Illustrating the History of Greek Mathematics: Vol. II Aristarchus to Pappus* London: Harvard University Press (1993) [The Loeb Classical Library].

66 The earliest reference to pi is contained in the Rhind papyrus (c. 1700 BCE) – "The area of a circle is equal to that of a square whose side is the diameter dimished by one-ninth." For example, for a circle with a diameter equal to 9, the area is (9-1) X (9-1) = 64. The actual area of the circle is calculated using the formula $A= \pi r^2 = 3.14 (4.5)^2 = 63.6$. Pi ($\pi$) is the ratio of the circumference of a circle to its diameter or $\pi$=C/d. Herman H. Harris, Jr. "The History and Calculation of Pi" *The Emporia State Research Studies*, Vol. 8, No. 1 (September 1959).

67 T. L. Heath (ed.) *The Works of Archimedes* New York: Dover Publications, Inc. "The Sand-Reckoner" [reprint of 1897 edition], p. 224.

Africa to India, a voyage which was not actually made before Vasco de Gama.[68]

Tarn also credits Eratosthenes with being the first to determine that it was possible to sail west from Spain to India.[69]

Eratosthenes used his extensive knowledge of mathematics to make an enormous contribution to Greek astronomy by accurately measuring the circumference of the earth. The calculation by Eratosthenes,

> …is based on simple observations made (at Alexandria) and Syene at the summer solstice, interpreted according to Euclid's propositions regarding equality of alternate angles and the similarity of arcs subtended by equal angles. [70]

Eratosthenes used a gnomon to measure the angle of the sun's rays that cast a shadow in Alexandria during summer solstice when the sun was directly overhead and casting no shadow in Syene which was located 5,000 stades to the south and almost along the same meridian (longitude).[71] Eratosthenes assumed no solar parallax, or that the sun's rays were parallel since the distance of the

---

[68] W. W. Tarn. *Hellenistic Civilization* 2nd ed. London: Edward Arnold & Co. (1930), p. 269. Tarn also cited Strabo that Eratosthenes believed that the inhabited earth was 8,910 miles X 4,340 miles divided latitudinally by the parallel of Rhodes (36°), p. 268.

[69] Ibid.

[70] P. M. Fraser. *Eratosthenes of Cyrene* London: Oxford University Press (1970), p. 16 [Proceedings of the British Academy, Vol. LVI]

[71] "The summer tropic, for instance, must pass through Syene, because there, at the time of the summer solstice, the index of the sun-dial does not cast a shadow at noon. And the meridian through Syene is drawn approximately along the course of the Nile from Meroe to Alexandria, and this distance is about ten thousand stadia; and Syene must lie in the centre of that distance…" Strabo. *The Geography of Strabo* 8 vols. (trans. by Horace Leonard Jones) London: William Heinemann Ltd. (1960) Vol. I, 2.5.7 (p. 439).

sun from the earth was very great. He observed that the angle of the sun's rays that cast a shadow in Alexandria was a part of a quadrant equivalent to 7.2 degrees. From this, he computed that the arc subtended was 1/50 of a circle (360 degrees/7.2 degrees = 50). Since the distance between Alexandria and Syene was known to be 5,000 stades, Eratosthenes determined that the circumference of the earth along the meridian that passed through both locations was 250,000 stades (50 X 5,000 stades = 250,000 stades). Eratosthenes later revised his computation of the circumference of the earth to 252,000 stades. [72]

Eratosthenes made several errors in his computations - Syene wasn't located on the Tropic of Cancer but some miles to the north, the distance between Alexandria and Syene was overestimated, and both locations were not on the same meridian. There are various estimates for the length of a stade but the calculation of the circumference of the earth by Eratosthenes was approximately correct. [73]

The earliest mention of a computation of the circumference of the earth is found in Aristotle's *On the Heavens*:

---

[72] Pliny. *Natural History*, ii, 108. Some commentators believe that Eratosthenes revised his figure for the circumference of the earth so that one degree equaled 700 stades. However, Hipparchus was the first to use degrees for angular measurement almost a century later.

[73] D. R. Dicks. *The Geographical Fragments of Hipparchus* London: University of London (1960), p. 42 "The Value of the Stade". According to Thomas L. Heath, Eratosthenes' calculation that the earth's circumference was 252,000 stades or 24,662 miles resulted in a polar diameter of 7,850 miles which is only 50 miles shorter than the actual polar diameter of the earth.

> Mathematicians who try to calculate the circumference [of the earth] put it at 400,000 stades.[74]

Aristotle lived more than a century before Eratosthenes. Nothing is known about the methods used by the earlier astronomers for making these calculations but, elsewhere in the same book, Aristotle discusses that the earth is spherical in shape and that the stars visibly change altitude as the observer changes location north or south.

> ...our observations of the stars make it evident, not only that the earth is circular, but also that it is a circle of no great size. For quite a small change of position to south or north causes a manifest alteration of the horizon. There is much change, I mean, in the stars which are overhead, and the stars seen are different, as one moves northward or southward. Indeed there are some stars in Egypt and in the neighborhood of Cyprus which are not seen in northerly regions; and stars, which in the north are never beyond the range of observation, in those regions rise and set.[75]

Poseidonius, much later, compared the altitude of the same star at different locations along the same meridian to calculate the circumference of the earth using a method similar to that used by Eratosthenes.

The geographer Strabo credits Eratosthenes with measuring the size of the inhabited world – the land masses comprising modern Europe, Africa and Asia – as having a length of 77,800 stadia and a width approximately half the length.

---

[74] Aristotle. *On the Heavens* (trans. by W. K. C. Guthrie) London: William Heinemann Ltd. (1939), Book II, XIV, 298a. A note states that, "This passage of Aristotle is said to have provided a stimulus to the voyage of Columbus (Ross, *Aristotle*, p. 96, n.3)."
[75] Aristotle. *De Caelo* Op. cit., Chapter 14 297b 30-298a 10.

At all events (Eratosthenes) says that the narrowest part of India up to the river Indus measures sixteen thousand stadia (for the part of India that extends to its capes will increase this length by three thousand stadia); and the distance thence to the Caspian Gates, fourteen thousand; then, to the Euphrates ten thousand, and from the Euphrates to the Nile five thousand, and on to its Canobic mouth thirteen thousand five hundred; then to the Pillars, at least eight thousand; there is accordingly, he says, an excess of eight hundred stadia over seventy thousand stadia. We must still add, he says, the bulge of Europe outside the Pillars, which lies over against Iberia and leans westward, reaching not less than three thousand stadia; we must also add all the capes...And he adds to the aforesaid length-distances still other stadia, namely, two thousand on the west, and two thousand on the east, in order to keep the breadth from being more than half the length.[76]

The latitude of cities and other geographic features was fixed by comparing the length of the shadow to the height of the sun-dial index at noon at the same time of the year in different locations.

Eratosthenes, according to Strabo, believed that it was possible to sail across the Atlantic Ocean from Iberia (Spain) west to India or from one side of the inhabited world to the other around the globe. However, he thought that this distance was nearly 200,000 stades.

"...in the manner of the mathematicians," (Eratosthenes) says, "(the inhabited world) forms a complete circle, itself meeting itself; so that, if the immensity of the Atlantic Sea did not prevent, we could sail from Iberia to India along one and the same parallel over the remainder of the circle, that is, the remainder when you have subtracted the aforesaid distance, which is more than a third of the whole circle – if it be true that the circle that runs through Athens, along which I have made the said reckoning of stadia from India to Iberia, is less than two hundred thousand stadia in circuit.[77]

---

[76] Strabo. *The Geography of Strabo* 8 vols. (trans. by Horace Leonard Jones) London: William Heinemann Ltd. (1960) Vol. I, 1.4.5 (pp. 239-241).

[77] Ibid., Vol. I, 1.4.6 (p. 241). A footnote explains:
"NOTE. It has been assumed by various scholars that Eratosthenes'

Eratosthenes is also credited with introducing the Babylonian sexagesimal system of mathematical notation to Greek astronomy.

The earliest evidence among Greek authors of sexigesimal units seems to be with Eratosthenes, i.e. around 250 B.C., in his division of the circumference of a circle in 60 parts...Aristarchus, some 50 years earlier, gives angular measurements in fractions of quadrants only; e.g. for an elongation of 87° = 90° – 3° he would say 'less than a quadrant by one-thirtieth of a quadrant.' [78]

This change had a profound influence on the development of Greek mathematical astronomy by simplifying measurements and calculations.

## HIPPARCHUS OF NICEA

Hipparchus of Nicea (194 BCE – 120 BCE) is sometimes known as the 'Father of Astronomy'. Like Eratosthenes, Hipparchus was a geographer who used astronomical observations to improve maps of the ancient world. Hipparchus developed mathematical geography using parallels (latitude) and meridians (longitude) to fix positions on the spherical earth. Earlier maps were drawn employing descriptive geography largely based on accounts of travels by soldiers, sailors, explorers, and merchants.

Hipparchus divided the circumference of the earth, that Eratosthenes had calculated, by the number of degrees in a sphere to compute the length of one degree at the equator: 252,000 stades/360 degrees = 700

---

parallel of latitude, above referred to, ran 25, 450 stadia north of the equator, which would be at 36º21'25½". In this case the circumference of this parallel works out to be 202,945 stadia – if we count 700 stadia to the degree, following Eratosthenes method..."

[78] O. Neugebauer. *A History of Ancient Mathematical Astronomy* New York: Springer-Verlag (1975) p.590.

stades/degree.[79] Hipparchus was the first astronomer to use degrees instead of parts of a quadrant to plot map locations. He also invented a method to determine meridian (longitude) by comparing times that the same lunar eclipse was visible at different locations.

Hipparchus, like Eratosthenes and others before him, used the ratio of the length of a gnomon (pointer of a sun dial) to the shadow it cast at equinox to determine parallel (latitude).[80] The geographical writings of Hipparchus are critical of Eratosthenes except where the latter used mathematical methods. Strabo (63 BCE – 21 CE), a geographer and commentator on Hipparchus, stated that,

> At all events it is a fact that many men have spoken of the necessity from wide learning in relation to this subject [i.e., geography]. Hipparchus also rightly points out in his treatise against Eratosthenes that, while geographical knowledge is the concern of everyone whether layman or scholar, it is impossible to attain it without consideration of the heavens and the observations of eclipses; thus one cannot determine whether Alexandria in Egypt is north or south of Babylon, or by how much, without investigation by means of the climata [latitude]. Similarly one cannot decide accurately whether places are situated to a greater or less degree towards the east or west

---

[79] D. R. Dicks. *The Geographical Fragments of Hipparchus* London: University of London (1960), pp.32-33. Hipparchus referred to bands of parallels (latitude) as 'climata' which is the origin of the word 'climate'.

[80] Equinox occurs in March and September each year when the equator and ecliptic meet. The ecliptic is a great circle formed by the transit of the sun through the constellations of the zodiac and it is inclined 23½° above the equator (known as the obliquity of the ecliptic). See E. G. Richards. *Mapping Time: The Calendar and its History* New York: Oxford University Press (1998), pp. 27-28. Comparison of the ratios of the height of gnomons to the length of their shadows allowed ancient mappers to determine the latitude of locations on the earth.

except by comparison of [the times of] eclipses of the sun and moon [longitude]. That is what Hipparchus says anyway. [81]

Hipparchus wrote *On the Sizes and Distances of the Sun and Moon.* He used principles of geometry (e.g., similar triangles) to calculate that the moon's radius was 1/1300 of the moon's orbit and that the distance of the moon from the earth was 67 times the earth's radius. He determined that the distance of the sun from the earth was 490 times the earth's radius based on an estimated horizontal solar parallax of approximately 7 minutes.[82]

The Roman encyclopedist Pliny the Elder commented on the catalog of stars compiled by Hipparchus which gave their names and fixed their positions and movements.

Hipparchus…detected a new star that came into existence during his lifetime; the movement of this star in its line of radiance led him to wonder whether this was a frequent occurrence, whether the stars that we think to be fixed are also in motion; and consequently he did a bold thing, that would be reprehensible even for a god – he dared to schedule stars for posterity, and tick off the heavenly bodies by name in a list, devising machinery by means of which to indicate their several positions and similar lines: which facts unquestionably demonstrate the power of nature, and prove that it is this that we mean by the word 'God'.[83]

---

[81] D. R. Dicks. *The Geographical Fragments of Hipparchus* London: University of London (1960), p. 65 Fragment 11 [my additions]. See also, M. Cary and E. H. Warmington. *The Ancient Explorers* Baltimore, MD: Penguin Books Rev. ed. (1963) [originally published 1929], p. 16.
"The fixing of longitudes remained throughout antiquity a mere matter of dead-reckoning. The astronomer Hipparchus, it is true, made the brilliant suggestion that the relative longitude of two places could be ascertained if observers at either place were to note the movement of first obscuration in a lunar eclipse moving from west to east; but this method was never applied in practice."

[82] Albert Van Helden. *Measuring the Universe: Cosmic Dimensions from Aristarchus to Halley*
Chicago, IL: University of Chicago (1985), pp. 11-13.

[83] Gaius Plinius Secundus (Pliny the Elder). *Natural History* (trans.

## POSEIDONIUS OF APAMEA

Poseidonius of Apamea (c. 130 BCE-50 BCE) calculated the circumference of the earth by comparing simultaneous observations of the altitude of the same star at two widely separated locations along the same meridian. Poseidonius found that the star Canopus lay on the horizon when observed at Rhodes while it had an altitude of 7-1/2 degrees when observed at Alexandria. The angle was 1/48 of a circle (360/7.5 = 48). Since the distance between Rhodes and Alexandria was known to be 5,000 stades, Poseidonius determined that the circumference of the earth along the meridian that passed through both locations was 240,000 stades (5,000 stades X 48 = 240,000 stades). [84]

Curiously, some scholars believe that the measurement of the circumference of the earth by Poseidonius influenced planning by Christopher Columbus for his voyage from Spain to India. Poseidonius was said to have visited Gades (modern Cadiz) and observed the ocean tides. He attributed the tides to changes in the phases of the moon.

> Someone, probably Eratosthenes, had suggested that the Atlantic might be divided longitudinally by land (i.e., America), a suggestion which inspired Seneca's famous prophecy of the discovery of a New World. Poseidonius, however, not only rejected this idea, but, having

---

By H. Rackham) 10 vols., Cambridge, MA: Harvard University Press (1967), Book II, 95 (p. 239) [Loeb Classical Library]

[84] Op. Cit., Fraser, p. 16. There is some disagreement whether the calculation by Poseidonius of the earth's circumference was actually a smaller value.
"That 240,000,...assigned to (Poseidonius), is only a later adjustment may be shown by his taking 140,000 as the circumference of the parallel of Rhodes." W. W. Tarn. *Hellenistic Civilization* 2nd ed. London: Edward Arnold & Co. (1930), p. 270, N.2.

got the size of the earth too small, believed that, on the parallel of Rhodes (36°), the 'inhabited world', which he made 70,000 stades across, was half the circumference; so, looking at the Atlantic, he very naturally remarked that a man sailing west for 70,000 stades would come to India. This remark, passed on by (among others) Roger Bacon, was the ultimate foundation of Columbus' confidence; and it was only historical justice that he should setsail for India from Poseidonius' Cadiz.[85]

Poseidonius computed that the distance from the earth to the moon was 125 times the radius of the earth (e.r.). For this calculation, Poseidonius assumed an apparent diameter for the moon equal to 1/750 part of its orbit and that the diameter of the moon was half the earth's diameter (or approximately 40,000 stades). As a result, Poseidonius determined that the lunar orbit was 750 X 40,000 stades or 30,000,000 stades and that the distance of the moon was the radius of its orbit or approximately 5,000,000 stades (125 e.r.). [86] Pliny, the early Roman encyclopedist, explained his calculations of the distances to the moon and sun in his *Natural History*:

Posidonius holds that mists and winds and clouds reach to a height of not less than 5 miles from the earth, but that from that point the air is

---

[85] W. W. Tarn. *Hellenistic Civilization* 2nd ed. London: Edward Arnold & Co. (1930), p. 271.

[86] Albert Van Helden. *Measuring the Universe: Cosmic Dimensions from Aristarchus to Halley* Chicago, IL: University of Chicago Press (1985), p. 14. Most ancient observers agreed that the apparent diameters of the sun and moon were approximately ½° or 1/720 of their orbits. Also, the earth's shadow on the moon during an eclipse suggested to them that the diameter of the earth was twice the diameter of the moon. Since the circumference of a circle is given as $C = 2\pi r$, then the radius of the lunar orbit can be calculated as $r = C \div 2\pi$ or $r = 30{,}000{,}000 \div 2\ (3) = 5{,}000{,}000$ stades. No similar conclusion was drawn comparing the diameter of the sun with the diameter of the earth. It was recognized that the moon had to be much nearer to the earth than the sun because the moon passed between the earth and sun during a solar eclipse but both the moon and sun had the same apparent diameter. The actual average distance from the earth to the moon is approximately 60 e.r.

clear and liquid and perfectly luminous, but that the distance between cloudy air and the moon is 250,000 miles and between the moon and the sun 625,000 miles, it being due to this distance that the sun's vast magnitude does not burn up the earth. [87]

The Stoics believed, as did Poseidonius, that the sun was fire and larger than the earth. Zeno of Citium (the Stoic) said the following:

The sun is pure fire: so Posidonius in the seventh book of his *Celestial Phenomena*. And it is larger than the earth, as the same author says in the sixth book of his *Physical Discourse*. Moreover, it is spherical in shape like the world itself according to this same author and his school. That it is fire is proved by its producing all the effects of fire; that it is larger than the earth by the fact that all the earth is illuminated by it; nay more, the heaven beside. The fact too that the earth casts a conical shadow proves that the sun is greater than it. And it is because of its great size that it is seen from every part of the earth.[88]

Poseidonius was the last of the important Greek astronomers that flourished before the common era. However, it should be noted that Claudius Ptolemaeus (Ptolemy) of Alexandria (c. 100 CE – 178 CE), a famous astronomer and geographer, authored the *Almagest* which described his 'geocentric' (earth-centered) model of the universe which dominated astronomy for the next 1,000 years. Ptolemy's work was a compilation of much of the product of the earlier Greek astronomers.

---

[87] Pliny. *Natural History* (trans. by H. Rackham) Vol. I London:William Heinemann Ltd. (1938), Book II, v. 85 (p.229).

[88] R. D. Hicks (trans.). *Diogenes Laertius: Lives of Eminent Philosophers* 2 vols. London: William Heinemann Ltd. (1965), p. 249, VII, Ch. 1, 144-145.

# CHAPTER 3
# THE BEGINNINGS OF RELIGIOUS UNBELIEF

> We may call the period from about 530 to 400 BC Greece's Age of Illumination, an age in which reason was striving to assert her rule in every sphere, and many superstitions, inherited from antiquity, were being challenged and discarded.[89]

The earliest civilizations often worshipped the sun, the earth, or animals because they sustained life. By contrast, the ancient Greeks explained the occurrence of natural events as the result of actions taken by an immortal race of anthropomorphic gods.

> These immortal beings, whose image has been impressed upon the world by aristocratic epic, are but faintly distinguished from the warrior heroes who worship them, save by the radiance of eternal youth. And sculptors, by investing them with a sovereign grace and a serene majesty, enabled them to elevate and ravish the souls of men by mere sight of their imperishable beauty. The whole spirit of the Hellenic religion, profoundly human, ideally aesthetic, as poets and artists had fashioned it, was opposed to the deification of celestial bodies, far-off powers, devoid of feeling and of plastic form.[90]

The poet Hesiod recounted ancient myths to describe the history of gods and mortal man in his *Theogony* (c. 700 BCE). Zeus, the son of the titan Cronos, emerged as the supreme god and he resided on Mount Olympus. Zeus employed lightning and thunder when angered. He ordered Hephaestus to form woman from clay and gave her a deceitful nature in order to punish man after the

---

[89] J. B. Bury, S. A. Cook, F. E. Adcock (eds.). *The Cambridge Ancient History* vol. V "Athens 478-401 BC" Cambridge University Press (1969), p. 376.

[90] Franz Cumont. *Astrology and Religion Among the Greeks and Romans* New York: G. P. Putnam's Sons (1912), p. 38.

titan Prometheus presented the use of fire to mortals. Zeus chained and punished Prometheus but later allowed the demigod Heracles to free him. Pandora, given as a gift from Zeus to the titan Epimetheus (brother of Prometheus) was the first woman. Pandora's curiosity caused plagues and evil to be released into the world when she opened a jar containing 'gifts' from the gods. Zeus destroyed man with a deluge but Deucalion (son of Prometheus) and his wife Pyrrha survived in a boat and landed on Mount Parnassus. They created a new race of people from stones they found.[91] According to Herodotus, sometimes known as the 'Father of History', Homer and Hesiod were the sources of the myths about the Greek gods.

> But it was only – if I may so put it – the day before yesterday that the Greeks came to know the origin and form of the various gods, and whether or not all of them had always existed; for Homer and Hesiod are the poets who composed our theogonies and described the gods for us, giving them all their appropriate titles, offices, and powers, and they lived, as I believe, not more than four hundred years ago. [92]

---

[91] There are many similarities between early Greek and Hebrew myths and the myths of other ancient civilizations such as the Babylonian or Egyptian. For instance, all these ancient civilizations had stories about great floods that the gods caused to punish mankind. The oldest is from Babylonia where the god Ea (or Enki) makes man from clay but later Enlil (king of gods) sends a flood to kill them. Ea warns Atrahasis in a dream to build a boat and take his family and animals to repopulate the world. See Yves Bonnefoy (ed.). *Mythologies* 2 vols. Chicago: The University of Chicago Press (1991), pp. 149-151. In ancient Greece, Zeus sent a flood and Prometheus warned Deucalion, his son, to take his wife Pyrrha and build a boat. Later they repopulated the world by throwing stones over their shoulders. The biblical story of Noah is well known and very similar. All three civilizations developed in the same region and shared myths and religious beliefs.

[92] Herodotus. *The Histories* (trans.by Aubrey de Selincourt) New York: Penguin Books (1972) Rev Ed.,
p. 151.

The poet Homer, originally from Ionia, was credited with compiling an authoritative history of the oral accounts of the war with Troy and its aftermath in about 750 BCE. His Iliad and Odyssey are thought to be among the earliest Greek writings. The roles of the gods in this history was accepted as the foundation of Greek religious mythology.

> The main difference between Greece and, say, Palestine is that the Greeks had no sacred writings comparable to the Old Testament, no definite dogmas to which subservience or at least acquiescence was required. The nearest thing to Scriptures were the Homeric poems, and these were definitely lay scriptures, not sacred ones. To be sure, Homer often referred to the gods, but that was incidental and he did it with a poet's license. Nevertheless, the Iliad and the Odyssey exerted a deep influence on Greek religion, for they helped to standardize the myths and to popularize them. Moreover, they humanized the gods and the heroes, sometimes to a degree that may scandalize the modern reader, but did not disturb the Greek listener. The latter knew that the gods were extremely powerful but he did not expect them to be perfect. Such as they were, Homer and Hesiod did not invent them but made them more familiar, consecrated their existence and their peculiarities.[93]

Some Greek philosophers, sophists (teachers) and poets disbelieved the myths about Olympian gods in the form of men as evidenced by fragments of their writings which have survived. Their careful observations of their world suggested natural causes for most events previously thought to result from supernatural forces. Not only did most of the sciences grow out of these beginnings, but some of the most revolutionary ideas of modern times had their roots in ancient Greece – evolutionary biology, atomic theory, and mathematical astronomy.

---

[93] George Sarton. *A history of Science: Ancient Science Throughout the Golden Age of Greece* Cambridge, MA: Harvard University Press (1952), p. 194.

The researches of scientific men in Ionia and in Magna Graecia had enlarged the range of knowledge and the number of subjects of which anyone aspiring to be a man of culture must have some knowledge, and thus a demand arose for higher education – for instruction in astronomy, geography, mathematics, physics, history. This demand was met by the sophists who were simply itinerant professors; collectively they performed the functions of a university in Greece.[94]

## XENOPHANES OF COLOPHON

Xenophanes of Colophon (c. 570-478 BCE) was one of the earliest Greek philosophers. Like other Greeks, Xenophanes believed in the gods but, unlike them, he disbelieved the myths as retold by Homer and Hesiod and he found them to be impious.

Unto the gods are ascribed by Hesiod, like as by Homer, all of the acts which are counted by men disgraceful and shameful, thieving and wenching and dealing deceitfully one with another. [95]

The writings of Xenophanes were the beginning of Greek unbelief about the mythology of popular religion but the writings were not irreligious. For example, Aristotle related the objections of Xenophanes to the birth of Olympian gods described by Hesiod.

It was a saying of Xenophanes that to assert that the gods had birth is as impious as to say that they die; the consequence of both is that there is a time when gods do not exist. [96]

---

[94] J. B. Bury, S. A. Cook, F. E. Adcock (eds.). *The Cambridge Ancient History* vol. V "Athens 478-401 BC" Cambridge University Press (1969), p. 377.

[95] Sextus Empiricus. *Against the Physicists* I, 193 (trans. by R. G. Bury) Vol. III Cambridge,MA: Harvard University Press (1968), p. 99.

[96] Aristotle. *Rhetoric* 1399b 5-8 in Richard McKeon (ed.) *The Basic Works of Aristotle* New York: Random House (1941), p. 1425.

Most notably, Xenophanes dismissed the anthropomorphic origins of the gods as purely an invention of man.

But mortals suppose that gods are born,
wear their own clothes and have a voice and body.

But if horses or oxen or lions had hands or could draw with their hands and accomplish such works as men,
horses would draw the figures of the gods as similar to horses, and the oxen as similar to oxen,
and they would make the bodies
of the sort which each of them had.

Ethiopians say that their gods are snub-nosed and black;
Thracians that theirs are blue-eyed and red-haired. [97]

Xenophanes was one of the first to question the truth of knowledge revealed to man by nature or by oracles. He said that,

No man knows truly anything and no man ever will. [98]

The religious criticism of Xenophanes influenced the thinking of Greek philosophers for centuries afterwards.

---

[97] J. H. Lesher (trans.). *Xenophanes of Colophon: Fragments* Toronto: University of Toronto Press (1992), p. 25, 85-94 Fragments 14-16. These fragments were published in *Stromateis* by Clement of Alexandria (c. 150-211 C.E.). This is an example of *reductio ad absurdum*. See, for example, H. Frankel. *Early Greek Poetry and Philosophy* (trans. by M. Hadas and J. Willis) New York (1973), pp.330-331 "…to render the anthropomorphic view wholly ridiculous, he crosses the boundary between man and beast."

[98] Diogenes Laertius. *Lives of the Eminent Philosophers* Vol. II (trans. by R. D. Hicks) Cambridge University Press (1965), "Pyrrho", B, ix , c. xi, 8.

## ANAXAGORAS OF CLAZOMENAE

Anaxagoras of Clazomenae (c. 500 BCE - 428 BCE), the first philosopher to live in Athens[99], was also the first Greek philosopher punished for impiety.

About this time also Aspasia (the mistress of Pericles) was put on trial for impiety, Hermippus the comic poet being her prosecutor... And Diopeithes brought in a bill providing for the public impeachment of such as did not believe in gods, or who taught doctrines regarding the heavens, directing suspicion against Pericles by means of Anaxagoras... Well then, Aspasia he begged off, by shedding copious tears at the trial, as Aeschines says, and by entreating the jurors; and he feared for Anaxagoras so much that he sent him away from the city.[100]

---

[99] Theodor Gomperz. *Greek Thinkers: A History of Ancient Philosophy* 2 vols. (trans. By Laurie Magnus) London: John Murray (1920) Vol. I, p. 208 [orig. publ. 1896].
Anaxagoras brought philosophy and natural science from Ionia to Attica.

[100] Bernadotte Perrin (trans.). *Plutarch's Lives* 10 vols. London: William Heinemann (1916), p. 93, XXXII, 1-3. Like his friend Anaxagoras, Pericles rejected supernatural causes for natural explanations as illustrated in the following anecdote (pp. 101, 103, XXXV, 1-2):

...But when the ships were already manned, and Pericles had gone aboard his own trireme, it chanced that the sun was eclipsed and darkness came on, and all were thoroughly frightened, looking upon it as a great portent. Accordingly, seeing that his steersman was timorous and utterly perplexed, Pericles held up his cloak before the man's eyes, and, thus covering them, asked him if he thought it anything dreadful. "No", said the steersman. "How then," said Pericles, "is yonder event different from this, except that it is something rather larger than my cloak, which has caused the obscurity?"

Historian Martin Nilsson commented on the implications of this anecdote as follows:

A solar eclipse was one of the greatest portents. If portents depended on natural causes, then they were not sent by the gods, and the art of divination was nothing but a phantom of the mind.

Martin P(ersson) Nilsson. *Greek Piety* (trans. By Herbert Jennings Rose) Oxford: Clarendon Press (1948), p. 80.

Later, Protagoras, Aristotle, and Epicurus would also flee Athens because of prosecution for impiety.

Anaxagoras taught that the sun and stars were burning stones and not gods.

> [The sun is] a red hot mass many times larger than the Peloponnesos.[101]

For this, Anaxagoras was banished from Athens and lived in exile at Lampsacos in Asia Minor. Like Xenophanes, Anaxagoras did not deny the existence of the gods but he imputed natural causes to astronomical phenomena. According to Martin Nilsson, historian of ancient Greek religion, Anaxagoras effectively undermined popular belief in eclipses and other astronomical events as omens or some other communication from the gods, when he advanced natural explanations instead. The result was that the seers opposed Anaxagoras and wanted him discredited.

> According to the biographer Satyros, Anaxagoras was accused by Thucydides, the son of Melesias, the chief political adversary of Pericles; but according to Plutarch, Diopeithes was the accuser. Probably the two worked together. Diopeithes carried a law in the popular assembly authorizing trials of persons who did not believe in the divine and who disseminated teachings about the celestial phenomena. Here we find the kernel of the matter, the clash between the old religion and the new philosophy [i.e., natural philosophy or science]. The heavenly bodies were mythological gods who had hardly any cult. The contention that the sun was a glowing lump and the moon another inhabited world could hardly be counted as atheism. On the other hand such celestial phenomena as eclipses had a very important place as omens in the art of seers. The seers became aware of the danger to their art, which some people had already begun to doubt, of the physical explanation of such phenomena.[102]

---

[101] Ibid., B ii, c. iii, 4.

[102] Martin P. Nilsson. *Greek Folk Religion* New York: Harper and Brothers (1961), p. 134. See also, Walter Burkert. *Greek Religion:*

This new thinking was rejected by the common man who was uneducated and superstitious. Plutarch, commenting on the actual cause of lunar eclipses first put forward by Anaxagoras, stated that,

For he who the first, and the most plainly of any, and with the greatest assurances committed to writing how the moon is enlightened and overshadowed, was Anaxagoras; and he was as yet but recent, nor was his argument much known, but was rather kept secret, passing only amongst a few, under some kind of caution and confidence. People would not then tolerate natural philosophers, and theorists, as they then called them, about things above; as lessening the divine power, by explaining away its agency into the operation of irrational causes and senseless forces acting by necessity, without anything of Providence or a free agent. [103]

The prosecution of Anaxagoras, like that of Socrates and Aristotle after him, was due more to local politics than his unconventional religious beliefs.

Whatever were the real causes of Anaxagoras' accusation [for impiety]- his friendship with Pericles or perhaps Persian leanings – the pretext was religious. Anaxagoras was indicted for rationalism

---

*Archaic and Classical* (trans. by John Raffan) Oxford: Basil Blackwell (1985), p. 316
But about 438(?) one Diopeithes, a seer, was able to carry through a novel decree of far-reaching consequences: one should 'denounce those who do not believe in the divine beings or teach doctrines about things in the sky.' It was known that this was directed against Anaxagoras and was meant to harm Pericles. Anaxagoras' most famous assertion was that Helios, like other heavenly bodies, is a glowing lump of metal, as the fall of a meteorite in 467 had proved to him. Now such doctrines were to be forbidden by the state. Anaxagoras had been teaching in Athens for 30 years; he now left the city. Thereafter, as far as we know, the decree was no longer used, and seems to have passed into oblivion. But the conflict between piety and natural science, indeed wisdom of all kinds, had come to stay; it re-echoes in the *Clouds* of Aristophanes as it does in Euripides.
[103] Plutarch. *The Lives of the Noble Grecians and Romans* Vol. I (Dryden translation revised by Arthur Hugh Clough) New York: The Modern Library (1992), p. 719 in "Nicias".

(c. 432). He was not the first victim in the incessant war between bigotry and science, but he is the first known one. [104]

Plato, because he subscribed to the popular belief that heavenly bodies were gods, and perhaps mindful of the persecution of his teacher Socrates on charges of impiety, condemned the views of Anaxagoras through the mouth of the Athenian character in *The Laws*.

It is rather the novel views of our modern scientists that we must hold responsible as the cause of mischief. For the result of the arguments of such people is this – that when you and I try to prove the existence of the gods by pointing to these very objects – sun, moon, stars, and earth – as instances of deity and divinity, people who have been converted by these scientists will assert that these things are simply earth and stone, incapable of paying any heed to human affairs. [105]

---

[104] George Sarton. *A History of Science: Ancient Science through the Golden Age of Greece* Cambridge, MA: Harvard University Press (1952), p. 244. See also, A. E. Taylor. "On the Date of the Trial of Anaxagoras" *Classical Quarterly* Vol. II, pp. 81-87 (1917). Taylor argues that Anaxagoras was prosecuted for "Medism" and impiety (stating that the sun was a burning stone) about the year 450 BCE – well before the Peloponnesian War – and that Anaxagoras fled to Lampsacus where he lived for many years and established a school. Taylor asserts that this early date for the trial explains why, "Anaxagoras is the only first-rate figure among the 'wits' of Athenian society in the Periclean age who Plato never represents Socrates as meeting."

[105] Plato. *The Laws* Vol. II (trans. by R. G. Bury) Cambridge, MA: Harvard University Press (1984), p. 303 [Book X]. Plato wrote that there were three kinds of impiety: disbelief in the existence of the gods, belief that the gods don't care about men, or belief that the gods can be bribed by offerings and prayers. Ibid., p. 299.

The 18th century British philosopher David Hume held a novel perspective on the alleged impiety of Anaxagoras. He argued that, since the heavenly bodies are merely burning rock or earth-like masses, then Anaxagoras would have been impious to worship them. In fact, his discovery implies the existence of a separate creator.

So, in effect, uses the argument of Anaxagoras' accusers against them.

It will be easy to give a reason, why Thales, Anaximander, and those early philosophers, who really were atheists, might be very orthodox

Here, Plato has framed the issue by describing the sources of the conflict between science and religion as rejection of supernatural causes and, therefore, denial of the existence of the gods. If the gods did not control nature and human affairs, then oracles, prayers, offerings, rituals, and other aspects of religion were meaningless.

> [A] conflict arose as soon as the heavenly phenomena, such as lightning and thunder, were ascribed to natural causes, or when the heavenly bodies were made out to be natural objects; for to the Greeks it was an established fact that Zeus sent lightning and thunder, and that the sun and moon were gods. A refusal to believe in the latter was especially dangerous because they were visible gods, and as to the person who did not believe in their divinity the obvious conclusion would be that he believed still less in the invisible gods. [106]

The naturalism of Anaxagoras had a profound impact on the subsequent development of science. It helped shape the way Greeks thought about their world and, ultimately, their religion.

---

in the pagan creed; and why Anaxagoras and Socrates, though real theists, must naturally, in ancient times, be esteemed impious. The blind, unguided powers of nature, if they could produce men, might also produce such beings as Jupiter and Neptune, who being the most powerful, intelligent existences in the world, would be proper objects of worship. But where a supreme intelligence, the first cause of all, is admitted, these capricious beings, if they exist at all, must appear very subordinate and dependent, and consequently be excluded from the rank of deities. Plato assigns this reason for the imputation thrown on Anaxagoras, namely, his denying the divinity of the stars, planets, and other created objects.
Quoted in Anthony Flew (ed.). *David Hume: Writings on Religion* La Salle, IL: Open Court (1992) – "The Natural History of Religion" p. 127. Reference to Plato is found in *The Laws*, X, 886A-E.

[106] A. B. Drachmann. *Atheism in Pagan Antiquity* Chicago, IL: Ares Publishers, Inc. (1977), p. 22. [reprint of the 1922 ed.].

## HIPPO OF RHEGIUM

Hippo of Rhegium (c.5th century BCE) was a Pythagorean and a contemporary of Anaxagoras. He was called "Hippo the Atheist" because he did not believe in the existence of the gods of mythology.

As to the reason why he earned the nickname *atheos*, it is said that he taught that water was the primal cause of all, as well as that he maintained that nothing existed but what could be perceived by the senses.[107]

According to Clement of Alexandria (fl.c. 150 CE), an early Christian apologist, Hippo believed that the gods of mythology were only men who later became worshipped as gods.

(Hippo) represented his death as a deification of himself. Thus Hippo ordered the following couplet to be inscribed on his monument:

*Behold the tomb of Hippo, whom in death*
*Fate made an equal of the immortal gods*

....This is the oracle of Hippo;let us understand its meaning. Those whom you worship were once men, who afterwards died. Legend and the lapse of time have given them their honours.[108]

The modern historian A. B. Drachmann believed this account of the inscription to be fictitious but explained its meaning similarly.

...Death has made (Hippo) the equal of the immortal gods (in that he now exists no more than they).[109]

---

107 Drachmann, op. cit., p. 29

108 G. W. Butterworth (trans.). *Clement of Alexandria* London: William Heinemann (1919) "The Exhortation to the Greeks", pp. 125-127.

109 Drachmann, op. cit., p. 30.

## HERACLITUS OF EPHESUS

Heraclitus of Ephesus (c. 540-480 BCE) was a generation behind Xenophanes but, like him, he had much to say about the popular religion which he considered impious. Where Xenophanes criticized religion for anthropomorphism, Heraclitus found idolatry.

> They purify themselves by staining themselves with other blood, as if one were to step into mud in order to wash off mud. But a man would be thought mad if any of his fellow-men should perceive him acting thus. Moreover, they talk to these statues (of theirs) as if one were to hold conversations with houses, in ignorance of the nature of both gods and heroes.[110]

Both Xenophanes and Heraclitus rejected popular conceptions about the nature of the gods but not the existence of the gods. For instance, Heraclitus believed that the gods were present in all things and animated them with a soul. He commented on a wide range of topics ranging from morality to the natural sciences but he was famous, even in his own era, for making cryptic remarks that often appeared contradictory or mysterious. For example, the meaning of his famous "river metaphor" has various interpretations among scholars.

> It is not possible to step twice into the same river.[111]

---

[110] Kathleen Freeman (trans.). *Ancilla to the Pre-Socratic Philosophers* Cambridge, MA: Harvard University Press (1957), p. 25. Fragment 5.

See also, Charles H. Kahn. *The Art and Thought of Heraclitus* London: Cambridge University Press (1979), p. 266. Kahn explains that the fragment refers to a blood sacrifice and the worship of idols. He says that pig's blood was used during purification rituals for blood guilt. Moreover, Kahn comments that,

...the target (of the criticism contained in this fragment) is not only the Apollonian rite of purification but the general practice of Greek religion as centered on temples and cult statues.

[111] Ibid., p. 31. Fragment 91.

Some believe that it refers to the nature of things as *becoming* and not *being* or, in other words, that things are not permanent but change constantly. Others speculate that the metaphor describes an individual's consciousness as *flowing* and his perception of things must also change.

Another example of a seemingly contradictory statement by Heraclitus was his assertion that, "good and ill are one." This is really a comment on the dual nature of things – a concept that was well known at that time. Moreover, it serves to help explain the existence of evil in a world created by the gods.

> This does not mean that good is evil or that evil is good, but simply that they are two inseparable halves of one and the same thing. A thing can be good only in so far as it is already evil, and evil only in so far as it is already good, and everything depends on the contrast…Just as the cessation of strife in the world would mean its destruction, so the disappearance of hunger, disease, and weariness would mean the disappearance of satisfaction, health, and rest.[112]

The puzzle of evil arises throughout the writings of the Greek philosophers, sophists and poets – and continues even up to this day.

## DIOGENES OF SINOPE

Diogenes of Sinope (c. 404 BCE – 323 BCE) was an early Greek naturalist whose ideas were ridiculed in the play *The Clouds* by the comic poet Aristophanes (445 BCE – 388 BCE). Aristophanes, who held conservative views, represented naturalism as a form of religious unbelief.

---

[112] John Burnet. *Early Greek Pjhilosophy* 4th ed. London: Adam & Charles Black (1945), p. 166. The fragment referred to is found in Diels (fragment 57).

Cicero, the great Roman writer (106 BCE – 43 BCE), asserted that Diogenes disbelieved in the existence of the gods because evil went unpunished.

Diogenes the Cynic used to say that Harpalus, who was regarded as one of the most successful robbers of his time, was a living witness against the gods, because he enjoyed so long a run of good luck. [113]

Arguments about the existence of evil have been propounded by skeptics of both pagan and modern religions. Some skeptics even rejected the idea that evil was ultimately punished in Hades. For example, Sextus Empiricus, referring to Homer's *Odysseus* argued that the dead couldn't suffer from punishments because they were not conscious.

For if Tityus was lifeless, how was he under punishment when he possessed no consciousness? And if he possessed life, how was he dead? … For if he never tasted drink or food how did he survive and not perish through lack of necessary sustenance? And if he was immortal, how is he in the state described? For an immortal nature is inconsistent with pains and torments, since everything that suffers pain is mortal. [114]

There is a widely reported anecdote regarding Diogenes that lends almost mythic stature to his fame while he was alive.

And now a general assembly of the Greeks was held at Isthmus, where a vote was passed to make an expedition against Persia with Alexander, and he was proclaimed their leader. Thereupon many statesmen and philosophers came to him with their congratulations, and he expected that Diogenes of Sinope also, who was tarrying in Corinth, would do likewise. But since that philosopher took not the slightest notice of Alexander, and continued to enjoy his leisure in the

---

[113] Cicero. *The Nature of the Gods* (trans. by Horace C.P. McGregor) London: Penguin Books (1972), Book 3, 82, p.229.
[114] Sextus Empiricus. *Against the Physicists* I, 68 (trans. by R. G. Bury) Vol. III Cambridge, MA: Harvard University Press (1968), p. 39.

suburb Craneion, Alexander went in person to see him; and he found him lying in the sun. Diogenes raised himself up a little when he saw so many persons coming towards him, and fixed his eyes upon Alexander. And that monarch addressed him with greetings, and asked if he wanted anything. "Yes," said Diogenes, "stand a little out of my sun." It is said that Alexander was so struck by this, and admired so much the haughtiness and grandeur of the man who had nothing but scorn for him, that he said to his followers, who were laughing and jesting about the philosopher as they went away, "But verily, if I were not Alexander, I would be Diogenes." [115]

## DIAGORAS OF MELOS

Diagoras of Melos (fl. c. 464 BCE), a pupil of Democritus, was notorious in ancient Greece as 'the atheist' because he questioned the existence of the gods. [116] In his play, *Birds*, Aristophanes mentions a decree issued at Athens in 415 BCE that condemned Diagoras for impiety and offered a reward for his capture or death. [117] The decree was said to have stated, "For anyone who kills Diagoras the Melian, a talent; for anyone who takes him alive, two talents."[118]

---

[115]Bernadotte Perrin (trans.). *Plutarch's Lives* 11 vols. London: William Heinemann Ltd. (1971) [Loeb Classical Library] Vol. VII *Alexander*, Chapter XIV, 2-5, p. 259.

[116] Ibid., Book 1, 2. Several religious skeptics in the ancient Greek world were labeled "the Atheist" because they questioned the truth of popular religious beliefs.

[117] Aristophanes "Birds" (trans. by R. H. Webb) in Moses Hadas (ed.) *The Complete Plays of Aristophanes* New York: Bantam Books (1962), p. 267

[118] See, Leonard Woodbury. "The Date and Atheism of Diagoras of Melos" *Phoenix* Vol. 19 (1965), p. 179:

"That the decree existed we know beyond a doubt, thanks to the Scholiasts on Aristophanes' *Birds* (1073) and *Frogs* (320). They tell us, on the authority of Melanthius' *On the Eleusinian Mysteries* that a proclamation was made at Athens, and its text inscribed in bronze, concerning Diagoras' denigration of the mysteries to the prejudice of initiations. Further, they state that Craterus, presumably in his *Collection of Decrees*, related a similar story, and add a quotation

The most common story about Diagoras of Melos illustrates how he challenged the existence of the gods with several arguments based on reason.

Diagoras, whom they call the atheist, visited Samothrace, where a friend remarked to him: "You believe that the gods are indifferent to human affairs but all these tablets with their portraits surely reveal to you the great number of those whose vows enabled them to escape the violence of a storm, so that they reached harbor safe and sound." "That is the case," rejoined Diagoras, "but there are no portraits in evidence of those who were shipwrecked and drowned at sea." This same Diagoras was once aboard ship when a storm blew up. The crew were fearful, and in their panic they told him that it served them right for allowing him to embark. He then pointed out to them several other vessels on the same route which were similarly in trouble, and he asked them whether they thought that there was a Diagoras aboard them as well. The fact is that one's character, and the kind of life which one has lived has no bearing on one's good or evil fortune. [119]

Diagoras was so infamous in ancient Greece for his unbelief that his name was used as an epithet to slander other religious skeptics. For instance, Aristophanes called Socrates 'the Melian' in his play *The Clouds* which was a transparent reference to Diagoras of Melos.[120]

---

taken from the text of the decree by Melanthius: 'for anyone who kills Diagoras the Melian, a talent; for anyone who takes him alive, two talents.' Craterus and Melanthius are good authorities of the third and fourth centuries B.C. and, because of their possessions of the text, are witnesses of the first importance."

[119] Cicero. *The Nature of the Gods*, Book 3, 89. Cotta is representing the Academy in the dialogue. This story, in shortened form, is attributed to Diogenes of Sinope by Diogenes Laertius. See his *Lives of Eminent Philosophers* (trans. by R. D. Hicks) Vol. II Cambridge University Press (1965), Book VI, 59.

[120] See Moses Hadas (ed.) *The Complete Plays of Aristophanes* New York: Bantam Books (1962), p.124.

Diagoras was said to have fled to Pellene, a city in Achaea that sided with Sparta against Athens in 431 BCE.

PROTAGORAS OF ABDERA

Protagoras of Abdera (c. 481 BCE – 411 BCE) was accused of impiety in Athens at approximately the same time that the decree was issued against Diagoras of Melos. Protagoras was a famous sophist, the first in Athens, and a contemporary of Socrates. Socrates quotes Protagoras as saying that,

> Man is the measure of all things, of the existence of the things that are and the non-existence of the things that are not. [121]

According to Socrates, this meant, "that individual things are for me such as they appear to me, and for you in turn such as they appear to you."[122] This shift of

---

[121] Plato. *Theaetetus*, Vol. VII, 152A (trans. by Harold North Fowler) Cambridge, MA Harvard University Press (1987), p. 41. Plato disagreed. In his *Laws*, the Athenian (protagonist) states: "What conduct, then, is dear to God and in his steps? One kind of conduct, expressed in one ancient phrase, namely, the 'like is dear to like' when it is moderate, whereas immoderate things are dear neither to one another nor to things moderate. In our eyes God will be 'the measure of all things' in the highest degree – a degree much higher than is any 'man' they talk of." Plato. *Laws* Vol. I, Book IV 716C (trans. by R.G. Bury) London: William Heinemann Ltd. (1967), p. 295. This was obviously a direct reference to the famous statement by Protagoras.
See also, Julia Annas and Jonathan Barnes (trans.) *Sextus Empiricus: Outlines of Scepticism* Cambridge University Press (1994), Book I, 219 (p.56):
"According to [Protagoras] then, man is the standard for what is; for all things that are apparent to men actually are, and what is apparent to nobody is not."

[122] Ibid. " In the Sophists' intellectual world, where nothing was accepted *a priori* anymore, the only sure criterion was immediate, concrete human experience. Gods, traditions, and mythical memories

reference from gods to men coincided with the growth of naturalism and the birth of science.

Plato and others record that Protagoras said that he didn't know anything about the gods.

Concerning the gods I am not able to say either whether they exist or of what sort they are; for the things which prevent me are many. [123]

This short statement was much debated during his lifetime and for many years afterwards. It demonstrates how far the opinions of the educated class had diverted from the traditional religious piety of the common man.

With these words he restricts the scope of his sentence about the impossibility of knowing the gods, and makes it an expression of an individual opinion. This must have seemed the only standpoint which would enable him to combine a fully precise approach to the question of objective certainty with his own personal recognition of the positive fact of religion and its undeniable significance for man as a social being. This approach must also have been extremely congenial to the tendencies of the time; for in spite of all the rationalistic criticism and all the current uncertainty, there was a growing feeling that religion was indispensable, and public opinion was becoming

---

no longer counted for anything. Our own judgements, our own feelings and interests now constituted the sole criteria. Jacqueline De Romilly. *The Great Sophists in Periclean Athens* (trans. by Janet Lloyd) Oxford: Clarendon Press (1992), p. 9.

[123] Sextus Empiricus. *Against the Physicists*, I, 56 (trans. by R. G. Bury) Cambridge, MA: Harvard University Press (1968), p. 33. Also, quoted in Diogenes Laertius. *Lives of the Eminent Philosophers* Vol. II (trans. by R. D. Hicks) Cambridge University Press (1965), Book IX, 51 (p.465). "In another work [*On the Gods*] he began thus: 'As to the Gods, I have no means of knowing either that they exist or that they do not exist. For many are the obstacles that impeded knowledge, both the obscurity of the question and the shortness of human life.' For this introduction to his book the Athenians expelled him; and they burnt his works in the market-place, after sending round a herald to collect them from all who had copies in their possession."

sensitive as never before to real and alleged attacks upon the gods of the state. [124]

Since Protagoras was a sophist, it is generally assumed that this fragment meant that the existence of the gods was an argument that could be proved or disproved. [125]

Protagoras denies the possibility of a sense-experience of the existence of the gods and of their non-existence; therefore it is in no way possible for him to find out what is the external manifestation of their being. This experience is prevented not only by the fact that the gods are not the object of sense-experience, but also by the brevity of human life which must renounce all attempts to argue by deduction from the perceptible to the imperceptible – an intellectual activity which remains in the realm of abstract possibility only. [126]

Even though considerable freethinking was tolerated in Athens during the 5th century BCE, questioning the existence of state gods was too much for the populace. Protagoras was eventually exiled from Athens.

## DEMOCRITUS OF ABDERA

Democritus of Abdera (c. 460 BCE-370 BCE), another pre-Socratic philosopher, was famous for teaching a mechanical world consisting of tiny atoms and the

---

[124] Werner Jaeger. *The Theology of the Early Greek Philosophers* Oxford: Clarendon Press (1947), pp. 189-90.

[125] Diogenes Laertius. *Lives of the Eminent Philosophers* Vol. II (trans. by R. D. Hicks) Cambridge University Press (1965), Book IX, 51 (pp.463-465). "Protagoras was the first to maintain that there are two sides to every question, opposed to each other, and he even argued in this fashion, being the first to do so."
Another perspective is provided by historian Hugh Lloyd-Jones in his *The Justice of Zeus* Berkeley, CA: University of California Press (1971), p. 131. "For Protagoras, the existence of Zeus was a matter not of knowledge, but of belief; but that does not make Protagoras an unbeliever."

[126] Mario Untersteiner. *The Sophists* (trans. by Kathleen Freeman) New York: Philosophical Library (1954), p. 27.

infinite void. Everything operated by natural causes which left no role for the gods.

Democritean cosmology...although not abandoning altogether the language of the divine, it implicitly denied (as Epicurus was to do explicitly) the view that the world is the product of design and purpose...Thus his naturalistic cosmology ...was unsettling and unorthodox, both in the sense that it was unconventional and in that it flew in the face of non-naturalistic notions of divine authorization that were deeply held and regularly and routinely acted upon in the Greeks' everyday religious practices. [127]

Democritus speculated on the origins of religious supernaturalism and concluded:

For when the men of old time beheld the disasters in the heavens, such as thunderings and lightnings, and thunderbolts and collisions between stars, and eclipses of sun and moon, they were affrighted, imagining the gods to be the causes of these things. [128]

In this thinking, Democritus was very similar to Anaxagoras. Fortunately, Democritus did not reside in Athens or he too may have been tried for impiety. Democritus, and Leucippus of Miletus [129], are important as the inventers of atomic physics but their remarkable ideas about invisible matter comprised the theoretical foundation of the materialist philosophy later propounded by Epicurus and his followers.

---

[127] Paul Cartledge. *Democritus* New York: Routledge (1999), p. 19

[128] Sextus Empiricus. *Against the Physicists*, I, 24 (trans. by R. G. Bury) Cambridge, MA: Harvard University Press (1968), p. 13.

[129] Cartledge, p.13

Who – and even whether – Leucippus was, is unclear today, as it seemingly was in antiquity too: Epicurus at any rate denied his very existence, and he was followed in that by at least one major commentator (Sextus Empiricus).

The importance that knowledge of the true nature of things held for the new scientists like Democritus is evident in one widely quoted comment attributed to him:

> (I would) rather discover one cause than gain the Kingdom of Persia.[130]

## CRITIAS OF ATHENS

Critias of Athens (480 –404 BCE)[131], a student of Socrates, an aristocrat and a member of the Thirty Tyrants that briefly ruled Athens in 404 BCE, is said to have written that man invented the gods to keep a civil society. A fragment of his play *Sisyphus* was preserved by Sextus Empiricus in *Against the Physicists*. Critias is considered, by historian A. B. Drachmann, to be the first unambiguous example of atheism or disbelief in the existence of the gods in ancient Greece.

A time there was when anarchy did rule
The lives of men, which then were like the beasts
Enslaved to force; nor was there then reward
For good men, nor for wicked punishment.
Next, as I deem, did men establish laws
For punishment, that Justice might be lord
Of all mankind, and Insolence enchained;
And whosoever did sin was penalized.
Next, as the laws did hold men back from deeds
Of open violence, but still such deeds
Were done in secret, - then, as I maintain,
Some shrewd man first, a man in counsel wise,
Discovered unto men the fear of gods,
Thereby to frighten sinners should they sin
Even secretly in deed, or word, or thought.
Hence was it that he brought in Deity,

---

[130] Kathleen Freeman. *Ancilla to the Pre-Socratic Philosophers* Cambridge, MA: Harvard University Press (1977), p. 104. Fragment 118.

[131] Not the same Critias that appears in several Socratic dialogues by Plato. See Burnet. *Early Greek Philosophy*, p. 203.

Telling how God enjoys an endless life,
Hears with his mind and sees, and taketh thought
An heeds things, and his nature is divine,
So that he hearkens to men's every word
And has the power to see men's every act.
Even if you plan in silence some ill deed,
The Gods will surely mark it; for in them
Wisdom resides. So, speaking words like these,
Most cunning doctrine did he introduce,
The truth concealing under speech untrue.
The place he spoke of as God's abode
Was that whereby he could affright men most, -
The place from which, he knew, both terrors came
And easements unto men of toilsome life-
To wit the vault above, wherein do dwell
The lightnings, he beheld, and awesome claps
Of thunder, and the starry face of heaven,
Fair-spangled by that cunning craftsman Time,-
Whence, too, the meteor's glowing mass doth speed
And liquid rain descends upon the earth.
Such were the fears where with he hedge men round,
And so to God he gave a fitting home,
By this his speech, and in a fitting place,
And thus extinguished lawlessness by laws...
Thus first did some man, as I deem, persuade
Men to suppose a race of Gods exists. [132]

Public expressions of impiety were not tolerated in Athens at this time so Critias must have been speaking through his character – a common practice employed by Euripides and other poets. However, the public understood this thinly veiled attempt to conceal the writer's beliefs.

---

[132] Sextus Empiricus. *Against the Physicists*, I, 54 (trans. by R. G. Bury) Cambridge, MA: Harvard University Press (1968), pp. 31-33. See also, Rosamond Kent Sprague (ed.) *The Older Sophists* Columbia, SC: University of South Carolina Press (1972), pp. 83. This book contains an English translation of the indispensable *Die Fragmente Der Vorsokratiker* edited by Diels-Kranz.

We have to remember that this speech was spoken by a character in a play, probably by that Sisyphus whose punishment by Zeus was a notorious example of divine justice; and however well the theory may accord with Critias' personal practice, he cannot have put it forward as his personal opinion.[133]

## PRODICUS OF CEOS

Prodicus of Ceos (fl. c. 470 BCE) was a famous sophist and a teacher of both Euripides and Diagoras of Melos. Little is known about Prodicus but he was noteworthy for claiming that the origins of religion were rooted in things found in nature that were useful to ancient man.

And Prodicus of Ceos says- "The ancients accounted as gods the sun and moon and rivers and springs and in general all things that are of benefit for our life, because of the benefit derived from them, even as the Egyptians deify the Nile." And he says it was for this reason that bread was worshipped as Demeter, and wine as Dionysus, and water as Poseidon, and fire as Hephaestus, and so on with each of the things that are good for use. [134]

Cicero (106 BCE – 43 BCE), the famous Roman writer and statesman, concluded that Prodicus was irreligious.

---

[133] Hugh Lloyd-Jones. *The Justice of Zeus* Berkeley, CA: University of California Press (1971), pp. 132-3.

[134] Sextus Empiricus. *Against the Physicists* (trans. by R. G. Bury) London: William Heinemann Ltd. (1936) Book I, 18, p. 11. See also, Sprague. *The Older Sophists*, pp.83-84. And see, Mario Untersteiner. *The Sophists* (trans. by Kathleen Freeman) New York: Philosophical Library (1954), p. 211:

...Prodicus tries to reconcile popular religion and philosophy in some fashion without annihilating in the supreme power of the God of Nature the separate gods who are in some way his manifestations. But this reconciliation does not mean that popular religion is actually preserved: the concept of progress, and cosmological pessimism, were bound in the last analysis to lead to the destruction of the traditional faith.

> Prodicus of Chios, who ascribed divinity to everything which benefits mankind: what room did he leave for religion? [135]

Modern historians disagree about whether his beliefs were an example of ancient atheism. Some think that there is clear evidence of atheism in a fragment contained in the book *On Piety* by Philodemus who commented on the religious ideas of the ancient Greek philosophers.

> [Prodicus] maintains that the gods of popular belief do not exist and that he does not recognize them, but that primitive man, [out of admiration, deified] the fruits of the earth and virtually everything that contributed to his subsistence…[136]

Historian A. B. Drachmann concluded that Prodicus did not actually consider things to be gods but that did not necessarily imply disbelief in the gods.

> As far as Prodicus is concerned it may be assumed that he did not believe that Bread, Wine, or Fire were gods, any more than Democritus imagined that Zeus sent thunder and lightning; nor presumably, did he ever believe that rivers were gods. But he need not therefore have denied the existence of Demeter, Dioysus and Hephaestus, much less the divinity of the sun and the moon. [137]

The thinking of Prodicus influenced many other writers and philosophers including Euhemerus and Theodorus – two well-known atheists in antiquity. There is some evidence that Prodicus may have been charged with impiety in Athens.[138]

---

[135] Marcus Tullius Cicero. *The Nature of the Gods* (trans. by Harace C. P. Mc Gregor) New York: Penguin Books (1972), p. 117

[136] Albert Henrichs. "Two Doxographical Notes: Democritus and Prodicus on Religion" *Harvard Studies in Classical Philology* vol. 79 (1975), pp. 107-108.

[137] A. B. Drachmann. *Atheism in Pagan Antiquity* Chicago, IL: Ares Publishers, Inc. (1977)[reprint of the original edition published in 1922], p. 43.

[138] Mary K. Lefkowitz. "'Impiety' and 'Atheism' in Euripides'

## BION OF BORYSTHENES

Bion of Borysthenes 'the Cynic' (c. 300 BCE) was a student of Theodorus of Cyrene who was eventually exiled from Athens for atheism. Theodorus was a contemporary of Socrates, a teacher of Plato, and he influenced Epicurus.

> Theodorus is one of the best-known atheists, who reject not only the popular gods but the idea of god in general. [139]

Bion was important in the history of ancient skepticism because he too rejected most forms of worship and the fundamental beliefs of popular religion. Even so, there is some disagreement by modern historians whether Bion was an atheist because the myths of Homer and Hesiod were disbelieved, or dismissed as allegories, by many in Greek society by his time. Also, there are too few fragments that have survived to definitively settle the issue.

> The extant fragments on religious questions clearly show that Bion has nothing positive to say on the subject...On the other hand it is not possible to assume from these fragments that Bion was an atheist...Yet Bion is highly critical of all the conventional forms and beliefs of popular religion. He does not accept prayers, the mysteries, the idea of punishment after death, or divination, he does not find fault with temple robbery and he challenges the idea that the gods are anthropomorphic and every form of superstition. [140]

---

Dramas" *Classical Quarterly* Vol. 39 (1989), p. 74.
"Ancient biographers believed that Euripides was a pupil of his younger contemporary sophist Prodicus, who like Anaxagoras and Protagoras was said to have been tried in Athens for impiety; according to the *Suda*, like Socrates, Prodicus 'died in Athens by drinking hemlock on the grounds that he corrupted the youth'."

[139] Jan Fredrik Kindstrand *Bion of Borysthenes: A Collection of the Fragments* Stockholm, Sweeden: Almquist & Wiksell Int'l. (1976), p. 69

[140] Ibid., p. 240.

Plutarch relates an anecdote about Bion and his criticism of the belief that the gods punished children for crimes committed by their parents.

Bion says that in punishing the children of the wicked God is more ludicrous than a physician administering medicine to a grandson or son for a grandfather's or father's disorder.[141]

## EURIPIDES OF SALAMIS

Euripides (480-406 BCE) was the author of numerous plays, many of which contain themes about the nature of the gods. In some of his tragedies, the characters even question the existence of the gods. Scholars disagree whether these characters reflect the views of Euripides on popular religion.[142]

---

See also, John M. Robertson. *A Short History of Freethought: Ancient and Modern* 2nd ed. 2 vols. London: Watts & Co. (1906) Vol. I,p. 180.
Among (Bion's) positions was a protest or rather a satire against the doctrine that the gods punished children for crimes of their fathers.

[141] Phillip H. De Lacy and Benedict Einarson. *Plutarch's Moralia* London: William Heinemann Ltd. (1959) "On the Delays of the Divine Vengeance" 19, 561C, p. 259.

[142] See, for example, Mary K. Lefkowitz. "'Impiety' and 'Atheism' in Euripides' Dramas" *Classical Quarterly* Vol. 39 (1989), p. 72, 75.
"But although Euripides' 'philosophizing' made him seem impious, at least to the comic poets and the biographers who used their works as 'evidence', I believe that it can be shown that any character in Euripides who expresses 'philosophical' notions about the gods does so out of desperation, and that ultimately, the gods in that play will prove – not always to the characters' satisfaction – that the gods still retain their traditional powers. Perhaps understandably, modern scholars often seem to be uncomfortable with such an austere notion of divinity." (p.72) "It is important to remember that to Euripides' contemporaries, honouring the gods did not mean liking them, or applauding their actions; on the contrary, the Athenian dramatists seemed concerned to portray them in all the emotions that the gods manifest toward mortals: cruelty, neglect, anger, loyalty, and

The whole purpose of some plays, such as the *Ion*, seems to be simply to discredit the gods. What are we to make of these contradictions? It is not as though they admitted of the ready solution, which would be obvious, did the abuse of the gods come only from the 'villains' of the plays and the praise from the 'heroes.' Such is not the case.[143]

However, some of his contemporaries, such as Aristophanes, considered Euripides to be an atheist. Perhaps the most often cited evidence of this finds Euripides speaking through his character in a fragment of *Bellerophon* which has survived.

Does someone say there are indeed gods in heaven? There are not, there are not if a man is willing not to rely foolishly on the antiquated reasoning. Consider for yourselves, do not base your opinion on words of mine. I say myself that tyranny kills very many men and deprives them of their possessions; and that tyrants break their oaths to ransack cities; and in doing this they are more prosperous under heaven than men who live quietly in reverence from day to day. I know too of small cities doing honour to the gods which are subject

---

occasionally, even affection, especially for one of their own children." (p.75)

See also, G.M.A. Grube. "Euripides and the Gods" in *Euripides: A Collection of Critical Essays* (ed. By Erich Segal) Englewood Cliffs, NJ: Prentice-Hall, Inc. (1968), pp. 42-43:

Euripides clearly aimed to represent life as he saw it. Actual human behavior was his starting-point and he was little concerned, in the first instance at least, with theological considerations. He paints men and women "as they are, not as they ought to be." Now one of the most deeply ingrained human characteristics is that in moments of overwhelming sorrow, when stunned by misfortune, especially if they feel the misfortune to be undeserved, men do doubt and (if they are pagans) curse their gods. But they do not doubt unless they believe, nor do they curse a god whom they know not to exist. It is not the atheist, but the believer who can be guilty of blasphemy.

[143] R. B. Appleton. *Euripides: The Idealist* London: J. M. Dent and Sons Ltd. (1927), pp. 166-167.

The author believed that Euripides wasn't concerned with the reality (existence) of the mythological gods but, instead, considered their nature.

to larger, more impious ones, because they are overcome by a more numerous army...[144]

Here, Euripides discusses a common theme in his plays that often evil goes unpunished by the gods. His character reasons that the gods don't exist because evil men often triumph over good men. One historian comments that the primary significance of Bellerophon is that it may reflect skepticism prevalent in Greek society at the time (c. 430 BCE).

It would never have occurred to Sophocles or Aeschylus to put such a speech in the mouth of one of his characters. When Euripides does that it is a proof that the question of the existence of the gods has begun to present itself to the popular consciousness at this time. [145]

---

[144] C. Collard, M. J. Cropp and K. H. Lee. *Euripides: Selected Fragmentary Plays* Vol. I
Warminster, England: Aris & Phillips Ltd. (1995), p. 105. Fragment 286 in "Bellerophon", pp. 98-120.
There is an ancient Mesopotamian poem (c. 1,000 BCE) in the form of a dialogue where one speaker complains about his bad fortune and the failure of the gods to help him despite his piety. See Yves Bonnefoy (ed.). *Mythologies* 2 vols. Chicago: The University of Chicago Press (1991), p. 169.
Stanza 7
In my youth, I devoted myself to the will of the gods.
With prostration and fervent prayer, I sought (them).
I wore as a yoke a servitude that brought me nothing in return.
My god assigned poverty to me instead of wealth.
A cripple is my superior, a lunatic outstrips me.
The rogue has been promoted but I have been brought low.
Those who neglect the gods go the way of prosperity,
While those who pray to the gods with fervor get nothing
But poverty and misery.
Stanza 8
The intentions of the gods [are as far from us] as the subsoil of heaven.
[It is not possible for men]
To understand them well.

[145] A. B. Drachmann. *Atheism in Pagan Antiquity* Chicago, IL: Ares Publishers Inc. (1977), p. 54.

The theme in *Bellerophon* about the triumph of evil over good is repeated in other plays. For example, in *Trojan Women*, the women of Troy lament the destruction of their city by the Greeks and wonder whether the gods are indifferent about their suffering. Homer's *Illiad*, which related the destruction of the city of Troy, provided the setting for several tragedies by Euripides. In each, the major characters raised questions about the existence of the gods.

Hecuba, queen of Troy, is grief-stricken and cries out these words when Talthybius, herald of the Greek King Agamemnon, notifies her of the sacrifice of her daughter Polyxena on the funeral pyre of Achilles:

What shall I say, Zeus? – that thou look'st on men?
On that this fancy false we vainly hold
For nought, who deem there is a race of Gods,
While chance controlleth all things among men?[146]

In *The Daughters of Troy*, Hecuba again calls out to Zeus and questions whether the god is actually just nature or an invention of man.

O Earth's Upbearer, thou whose throne is Earth,
Whoe'er thou be, O past our finding out,
Zeus, be thou Nature's Law, or Mind of Man,
Thee I invoke; for, treading soundless paths,
To Justice' goal thou bring'st all mortal things.[147]

---

[146] Arthur S. Way (trans.) *Euripides* Vol. I London: William Heinemann (1920), "Hecuba", 488-491 (p. 287)

[147] Ibid., "The Daughters of Troy", 884-888 (p. 427). See also:
W. K. C. Guthrie. *The Greeks and Their Gods* Boston: Beacon Press (1954), p. 142
O Thou who at once art prop and stay of the Earth and broodest over it, whosoever thou art, hard to divine or know – Zeus, be thou compelling force of nature or the mind of man, to thee I pray. For treading thy noiseless way thou leadest aright all the things of mortals. Werner Jaeger. *The Theology of the Early Greek Philosophers* Oxford: Clarendon Press (1947), p. 243 [Note 59].

The Greeks had similar questions about the gods provoked by tragedies of their own. *Iphigenia in Tauris*, is a play about the unlikely reunion of Iphigenia, eldest daughter of Agamemnon, King of Argos, with her brother Orestes. Iphigenia had been given by Agamemnon for sacrifice as demanded by an oracle her father consulted about favorable weather for the Greek fleet to sail to Troy. At the last moment, the goddess Artemis saved Iphigenia and made her a temple priestess at Tauris. Her brother Orestes finds the acts of the gods incomprehensible and rejects the explanations of the priests.

Aye; the gods too, whom mortals deem so wise,
Are nothing clearer than some winged dream;
And all their ways, like man's ways, but a stream
Of turmoil. He who cares to suffer least,
Not blind, as fools are blinded, by a priest,
Goes straight… to what death, those who know him know.[148]

Iphigenia defends the gods against stories in the myths which depict them as given to evil deeds. She believes that men are simply projecting their own depraved nature onto the gods.

---

Thou, earth's support, enthroned on earth, whoever thou mayest be, hard to discover, Zeus, be thou nature's law or the mind of man, I pray to thee: for by a noiseless course thou guidest human fate in righteousness.
Walter Burkert. *Greek Religion: Archaic and Classical* (trans. by John Raffan) Oxford: Basil Blackwell Ltd (1985), p. 319
In his Trojan Women, [Euripides] has Hecabe pray:
You who bear the earth and repose on it, whoever you are, difficult to guess and to know, necessity of nature or mind of men – I pray to you: moving in soundless ways you lead the affairs of mortals in accordance with justice.
[148] Lane Cooper (ed.) *Fifteen Greek Plays* New York: Oxford University Press (1956), 10th printing, "Iphigenia in Tauris" (trans. by Gilbert Murray), p. 417.

These tales be lies, false as those feastings wild
Of Tantalus and Gods that tore a child.
This land of murderers to its god hath given
Its own lust; evil dwelleth not in heaven.[149]

Euripides even questions whether death may be the only refuge from the world's evil.

Who knoweth if the thing that we call death
Be Life, and our Life dying – who knoweth?
Save only that all we beneath the sun
Are sick and suffering, and those forgone
Not sick, not touched with evil any more.[150]

In *Electra*, Orestes conspires with his sister to kill their mother, Clytemnestra, and her husband, Aegisthus, in revenge for murdering their father, Agamemnon, when he returned from Troy. Orestes swears to protect his sister and get justice for them both. He claims that, if Clytemnestra and Aegisthus aren't punished, it would be proof that there are no gods.

Thy defender, yea alone
To fight the world! Lo, this day I have thrown
A net, which once unbroken from the sea
Drawn home, shall… O, and it must surely be!
Else men shall know there is no God, no light
In heaven, if wrong to the end shall conquer right.[151]

Euripides questions whether there is even a heaven or an afterlife.

---

[149] Ibid., p. 412. Also, see Lewis Richard Farnell. *Outline-History of Greek Religion* Chicago: Ares Publishers (1974), p. 115.
[Euripides] leading principle of criticism …is expressed in the Iphigeneia in Tauris, namely that the evil in religious practice and legend arises from men imputing their own evil nature to God.
[150] Gilbert Murray. *Euripides and His Age* New York: Henry Holt and Co. (1913), p. 190.
[151] Ibid., p. 369.

And whatever far-off state there be,
Dearer than Life to mortality,
The hand of the Dark hath hold thereof
And mist is under and mist above.[152]

In *Zeus Rants*, the satirist Lucian speaks through the character Damis (an Epicurean like Lucian) who is debating the existence of the gods and comments on some controversial lines from Euripides.

Damis. I assure you when Euripides, following his own devices, says what he thinks without being under any constraint imposed by the requirements of his plays, you will hear him speaking frankly then:
*Dost see on high this boundless sweep of air*
*That lappeth earth about in yielding arms?*
*Hold this to be Zeus, and believe it God.*
And again:
*Twas Zeus, whoever Zeus is, for I know Him not, except by hearsay.*[153]

Plutarch, in his *Moralia*, comments on the controversy that caused Euripides to change the lines at the beginning of his (now lost) play *Melanippe the Wise*:

You have no doubt heard what an uproar burst upon Euripides when he began his *Melanippe* with this verse:
*Zeus, whoever he is, for I know him only by report.*
Well, he got another chorus (for he had confidence in the play, it seems, since it was composed in an elevated and elaborate style) and changed the verse to the present text:
*Zeus, as the voice of truth declares.*[154]

---

[152] Gilbert Murray. *Euripides and His Age* New York: Henry Holt and Co. (1913), p. 191.
[153] A. M. Harmon (trans.) *Lucian* 8 vols. London: William Heinemann Ltd. (1953) Vol. II "Zeus Rants", p. 153 [Loeb Classical Library]. The first quote is from a lost play and included by Cicero in *De Natura Deorum*, II, 25 (p. 187), H. Rackham (trans.), Cambridge, MA: Harvard University Press (2000). The second quote is widely cited as contained in the lost play *Melanippe the Wise*.
[154] Edwin L. Minar Jr., F. H. Sandbach, W. C. Helmbold (trans.) "The Dialogue on Love" in *Plutarch's Moralia* 15 vols. London:

There is an account that Euripides was prosecuted for impiety.

> Every one disliked [Euripides], the men because of his unsociableness, the women because of the censures in his poems. And he incurred great danger from both sexes, for he was prosecuted by Cleon the demagogue in the action for impiety...[155]

The charge was brought against Euripides by Cleon, a well-known Athenian politician.

> The accuser of Euripides was no less a person than Cleon the politician. But the prosecution seems to have failed; nothing happened to interfere with the usual activities of the poet. It is quite possible that Cleon was a man of very orthodox beliefs, who hated the advanced views of Euripides and considered them dangerous to society...It is however quite possible that the charge of blasphemy may have been rather a pretext than the true reason, which may have been the desire of silencing an influential critic who had access to the ears of the public, and whose ideas of justice and humanity were outraged by some political acts of Cleon, as they assuredly must have been by his proposal to put to death the population of Mitylene.[156]

Gilbert Murray, the incomparable translator of Euripides[157], probably understood Euripides and his times best. He summarized the prosecution of Euripides for impiety at Athens and put it in the proper perspective.

---

William Heinemann Ltd. (1961), 756B-C (pp. 347-349). See also, D. L. Page (trans.). *Greek Literary Papyri* 2 vols. London: William Heinemann Ltd. (1942) Vol. I, "Melanippe the Wise", pp. 117-119.

[155] Arthur S. Hunt (trans.) *The Oxyrhynchus Papyri* Part IX London: Egypt Exploration Fund (1912), p. 177 From, Satyrus (biographer). *Life of Euripides* Papyrus #1176, Fragment 39 (from 6th book of 'Lives' of Satyrus) Only fragments have survived.

[156] J. B. Bury, S. A. Cook, F. E. Adcock (eds.) *The Cambridge Ancient History* Vol. V "Athens 478-401 B.C." Cambridge University Press (1969), p. 383.

[157] In my opinion, Gilbert Murray's translation of Euripides' plays into English verse is reminiscent of Edward FitzGerald's famous translations of *The Rubaiyat of Omar Khayyam.*

The difference of atmosphere between the sophists of the Periclean Circle and the ordinary backward Attic farmer must have been visible to every observer. If more evidence of the great gulf was needed, it was supplied emphatically enough in the experience of Euripides. He himself was prosecuted by Cleon, the demagogue, for "impiety." The same charge had been leveled even against his far less destructive predecessor, Aeschylus. Of (his)...special friends...Euripides did not live to see Socrates condemned to death and executed. But he saw Anaxagoras, in spite of the protection of Pericles, accused of "impiety" and compelled to fly for his life. He saw Protagoras, for the book which he had read aloud in Euripides' own house, prosecuted and condemned. The book was publicly burned; the author escaped, it is said, only to be drowned at sea, a signal mark in the eyes of the orthodox of how the gods regarded such philosophy.

Thought was no doubt freer in ancient Athens than in any other city within two thousand years of it. Those who suffered for religious advance are exceedingly few. But it was not in human nature, especially in such early times, for individuals to do such great service for their fellow men and not occasionally be punished for it. They induced men for a time to set reason and high ideals above the instincts of the herd: and sooner or later the herd must turn and trample them.[158]

Clearly, the conflict between science and religion in ancient Greece had few victims but the conflict was present nevertheless. Athens was the site of the first skirmishes in the conflict. More than just science and religion were opposed, as would be the case much later with Galileo and others. Also, involved were politics, class, and culture.

Euripides, possibly because of his portrayal of the gods, did not win many prizes for his plays in Athens. However, there is evidence that his plays were popular elsewhere in the Greek world and, in time, it is arguable that he became the most famous Greek poet. For example, after the disastrous Sicilian expedition, many

---

[158] Gilbert Murray. *Euripides and His Age* New York: Henry Holt and Co. (1913), pp.57-58.

Athenian prisoners reported being freed upon reciting lines from Euripides' plays.

Some [Athenian soldiers taken prisoner] also were saved for the sake of Euripides. For the Sicilians, it would seem, more than any other Hellenes outside the home land, had a yearning fondness for his poetry. They were forever learning by heart the little specimens and morsels of it which visitors brought them from time to time, and imparting them to one another with fond delight. In the present case, at any rate, they say that many Athenians who reached home in safety greeted Euripides with affectionate hearts, and recounted to him, some that they had been set free from slavery for rehearsing what they remembered of his works…[159]

The impact of Euripides on his age was profound as evidenced by his circle of friends in Athens and by the criticism of him in many of the writings produced then and afterwards. For example, the popular comic poet Aristophanes used his wit to attack the progressive and, it was alleged, atheistic thinking present in the plays of Euripides.

Of the eleven comedies of Aristophanes which have come down to us three are largely devoted to Euripides, and not one has managed altogether to avoid touching him. I know of no parallel to it in all the history of literature….the hostility of the comic writers had evidently a general hostility behind it. Our tradition states this definitely and the persistency of the attacks proves it.[160]

The views of popular historian, Will Durant, on the nature of the writings of Euripides, probably best represents the consensus of opinion expressed by most scholars, excepting some few modern historians who may be ignoring the surviving writings of persons who lived in ancient Greece.

---

[159] Bernadotte Perrin (trans.). *Plutarch's Lives* 10 vols. London: William Heinemann Ltd. (1932) [Loeb Classical Library] Vol. XXIX "Nicias", p. 309.

[160] Gilbert Murray. *Euripides and His Age* New York: Henry Holt and Co. (1913), pp. 30, 31.

But it would be foolish to judge Euripides chiefly as a playwright; his ruling interest is not dramatic technique but philosophical inquiry and political reform. He is the son of the Sophists, the poet of the Enlightenment, the representative of the radical younger generation that laughed at the old myths, flirted with socialism, and called for a new social order in which there should be less exploitation of man by man, of women by men, and of all by the state. It is for these rebel souls that Euripides writes; for them he adds his skeptical innuendoes, and inserts a thousand heresies between the lines of supposedly religious plays. He covers his tracks with pious passages and patriotic odes; he presents a sacred myth so literally that its absurdity is manifest and yet his orthodoxy cannot be impeached; he gives the body of his plays over to doubt, but surrenders the first and last words to the gods. His subtlety and brilliance, like those of the French Encyclopedists is due in some part to the compulsion laid upon him to speak his mind while saving his skin. His theme is that of Lucretius – *Tantum religio potuit suadere malorum* so great are the evils to which religion has led men: oracles that breed violence upon violence, myths that exalt immorality with divine example, and shed supernatural sanctions upon dishonesty, adultery, theft, human sacrifice, and war. He describes a soothsayer as "a man who speaks few truths but many lies"; he calls it "sheer folly" to chart the future from the entrails of birds; he denounces the whole apparatus of oracles and divination.[161]

## SOCRATES OF ATHENS

Socrates of Athens (468 BCE – 399 BCE) was one of the most influential ancient Greek philosophers although he left no writings. He was also famous because of his persecution for impiety at Athens. Socrates, by then an old man, was found guilty, imprisoned and forced to drink poison. Socrates was a student of Archelaos of Miletus who learned from Anaxagoras of Clazomenae. Socrates was well known in his time and many considered him to be a sophist like Protagoras.

---

[161] Will Durant. *The Life of Greece* New York: Simon and Schuster (1966), p. 413.

Aristophanes (445 BCE – 388 BCE) portrayed Socrates as irreligious in his play *The Clouds:*

Socrates. What Zeus? Don't be silly; there is no Zeus.
Strepsiades. No? Then who rains. First tell me that.
Socrates. These clouds, of course, and I'll prove it by evidence plain. Come, did you ever see rain without clouds? But Zeus should bring rain from a clear sky, when clouds are vacationing…
Strepsiades. But the lightning – tell me whence come its blazing flashes which char or singe us. Surely Zeus firing at perjurers?
Socrates. Fool, you reek of antediluvian ignorance. If perjurers are the target why has Simon never been struck, or Cleonymus, or Theorus, perjurers all? But Zeus' own temple he strikes. And Sunium, Athens' headland, and the mighty oaks why? Oaks are no perjurers. [162]

Aristophanes also portrayed Socrates as a proponent of the new science in *The Clouds* by depicting Socrates explaining a natural cause for lightning instead of attributing it to Zeus.

The discussions of the Sophists were beyond the horizon of the common people, who listened to them partly in amusement and partly in irritation. It is characteristic that Euripides, the spokesman of the new wisdom on the stage, won few victories, while many fell to Sophocles. Sophocles won the favor of the people because he was a good Athenian citizen who believed in the gods…The intellectual arguments of the Sophists were above the understanding of the common people. The arguments of natural philosophy, at least to a certain degree, were not. Aristophanes popularized them. In *The Clouds* he makes Socrates prove that Zeus does not exist by the fact

---

[162] Moses Hadas (ed.) *The Complete Plays of* Aristophanes New York: Bantam Books (1988) "Clouds" (trans. By Moses Hadas), pp. 112-113. See also, Jeffrey Henderson (trans.) *Aristophanes Vol. II* "The Clouds" l. 367-370 Cambridge, MA: Harvard University Press (1998). See also, Leo Strauss. *Socrates and Aristophanes* New York: Basic Books Inc. (1966), p. 19. In addition, see Gilbert Murray. *Aristophanes: A Study* Oxford: Clarendon Press (1933), p. 89.
Clouds are deities of the sect Socrates worships – they, not Zeus, make it rain.

that the thunderbolt hits not the wrongdoers but the temples, mountaintops, and tall oaks. This the people understood.[163]

Socrates was not an atheist, in that he didn't deny the existence of the gods, but he clearly didn't believe in the popular religion. For the most part, the ideas of Socrates are reported by his pupil Plato. In Plato's *Republic*, Socrates addresses the question of how evil can exist if the gods are good:

Socrates. And no good thing is hurtful?
Adeimantus. No, indeed.
Socrates. And that which is not hurtful hurts not?
Adeimantus. Certainly not.
Socrates. And that which hurts not does no evil?
Adeimantus. No.
Socrates. And can that which does no evil be a cause of evil?
Adeimantus. Impossible.
Socrates. And the good is advantageous?
Adeimantus. Yes.
Socrates. And therefore the cause of well-being?
Adeimantus. Yes.
Socrates. It follows therefore that the good is not the cause of all things, but of the good only?
Adeimantus. Assuredly.
Socrates. Then god, if he be good, is not the author of all things, as the many assert, but he is the cause of a few things only, and not of most things that occur to men. For few are the goods of human life, and many are the evils, and the good is to be attributed to god alone; of the evils the causes are to be sought elsewhere, and not in him.[164]

Socrates logically argues that the gods are good and therefore can only be responsible for the good that exists. This argument raises new questions however. What is the source of evil? Are the gods not all powerful? These questions will likewise confront Epicurus and his followers and, later, the Christian

---

[163] Martin P. Nilsson. *Greek Folk Religion* New York: Harper and Brothers (1961), pp. 135-136.
[164] Plato. *Republic* (trans. by B. Jowett) New York: Modern Library (1982), Book II, p.75.

apologists after the pagan religions have been all but forgotten.

Socrates did not speak to religion directly- or even talk about the nature of the universe - but, instead, was concerned with ethics and man. The historian Xenophon wrote a defense of Socrates, albeit after his death, and argued that Socrates did not teach religion or natural sciences.

> But no one ever saw Socrates doing, or heard him saying, anything impious or unholy. For he did not converse about the nature of all things in the way most of the others did – examining what the sophists call the cosmos: how it is, and which necessities are responsible for coming to be of each of the heavenly things. But he even showed that those who worry about things of this sort are foolish. [165]

Much has been written about the trial and death of Socrates. It is often considered an early example of religious intolerance. Socrates was accused by three fellow Athenians – Meletus, Anytus and Lycon – of impiety in that he corrupted the youth by teaching them not to believe in the state gods. During the trial, Socrates cross-examined Meletus about the charges:

> Socrates. Do you mean that I teach the young to believe in some gods, but not in the gods of the state? Do you accuse me of teaching them to believe in strange gods? If that is your meaning, I myself believe in some gods, and my crime is not that of complete atheism. Or do you mean that I do not believe in the gods at all myself, and that I teach other people not to believe in them either?
> Meletus. I mean that you do not believe in the gods in any way whatever.
> Socrates. You amaze me Meletus! Why do you say that? Do you mean that I believe neither the sun nor the moon to be gods, like other men?

---

[165] Xenophon. *Memorabilia* (trans. by Amy L. Bonnette) Ithaca, NY: Cornell University Press (1994), Book I, Chapter 1, 11 (p.3).

Meletus. I swear he does not, judges. He says that the sun is stone and the moon earth.
Socrates. My dear Meletus, do you think you are prosecuting Anaxagoras? You must have a very poor opinion of these men, and think them illiterate, if you imagine that they do not know that the works of Anaxagoras of Clazomenae are full of these doctrines. [166]

Socrates denied that he was an atheist but was eventually found guilty by a vote of 281 to 220 and condemned to death. Socrates refused to leave Athens and cease his teaching because he believed that, "an unexamined life is not worth living." [167] Later, Socrates died in prison after drinking poison.

Plato's writings on Socrates are concerned with ethics – not religion. However, some historians believe that Socrates must have been widely perceived as a sophist, and therefore an atheist, because of the manner in which he was depicted in the comedy *The Clouds* by Aristophanes. [168]

---

[166] Plato. *Apology*, XIV, in *Euthyphro, Apology and Crito and Death Scene from Phaedo* (trans. by F. J. Church) New York: The Liberal Arts Press (1956), 2[nd] rev. ed., p. 32

[167] Ibid., XXVIII, p. 45.

[168] A. B. Drachmann, *Atheism in Pagan Antiquity* Chicago, IL: Ares Publishers Inc. (1977), p.65.
Drachmann succinctly describes the general opinion about sophists as represented by Aristophanes in *The Clouds*: "Here we find all the features of the[sophistic] school grotesquely mixed up and distorted by the farce, it is true, but nevertheless easily recognizable: rhetoric as an end in itself, of course, with emphasis on its immoral aspect; empty and hair-splitting dialectics; linguistic researches; Ionic naturalism; and first and last, as the focus of all, denial of the gods." (p. 57)
It must be understood that the common man was not greatly influenced by the underlying criticism of the gods contained in many plays.
The occasional fun poked at the gods in comedy is not evidence against the religious conservatism of the ordinary man; it is when a religion is sure of itself that such amusement is permitted.

Most historians of the period argue that the trial of Socrates for impiety was really an attack on a proponent of the aristocracy by his political enemies in the democracy that returned to power at Athens just five years previously in 404 BCE.[169] A similar fate awaited Aristotle, the pupil of Plato.

> But the predominate feeling of the Athenian people is clearly manifested by the circumstance that, after the lapse of more than half a century, the statesman and orator Aeschines could hope to advance the cause which he was then promoting by addressing the assembled people in the following words: "Again, men of Athens, you put to death Socrates the sophist, because it was proved that Critias, one of the thirty destroyers of the democracy, had been educated by him.[170]

---

Hugh Lloyd-Jones. *The Justice of Zeus* Berkeley, CA: University of California Press (1971), p. 133.

[169] Martin P. Nilsson. *A History of Greek Religion* (trans. by F. J. Fielden) 2nd ed., New York: W. W. Norton & Company Inc. (1964), p. 275. "The persecution of Socrates was one of the first acts of the restored democracy after the expulsion of the thirty tyrants."

See also, Diels-Kranz (ed.) *The Older Sophists* (trans. by Rosamond Kent Sprague) Columbia, SC: University of South Carolina Press (1972), 88 "Critias" (trans. by Donald Norman Levin) (p. 241):

Both (Critias) and Alcibiades, similarly brilliant and unscrupulous, spent much time in Socrates' company. Probably the reckless behavior of both had not a little to do with the later charge that Socrates was guilty of corrupting the young men of Athens – this despite the fact that the sage himself had refused to countenance either Alcibiades' treachery or Critias' outrageous behavior as a member of the oligarchical Thirty.

[170] Theodor Gomperz. *Greek Thinkers: A History of Ancient Philosophy* 2 vols. (trans. By G. G. Berry) New York: Scribner's Sons (1905) Vol. II, p. 118.

See also, Charles Darwin Adams (trans.) *The Speeches of Aeschines* London: William Heinemann (1919), p. 139 "Against Timarchus" 1. 173 [Loeb Classical Library].

Did you put to death Socrates the sophist, fellow citizens, because he was shown to have been the teacher of Critias, one of the Thirty who put down the democracy…

Xenophon gave the following account of the nature of the charges made against Socrates in his *Memorabilia*:

> "But, by Zeus," the accuser said, "(Socrates) made his companions look down upon the established laws, by saying that it is foolish for the rulers of the city to be established in office by lot, when nobody would want to use a pilot chosen by lot or a builder or a flutist or anyone for any other task of the sort, although far less harm is caused when mistakes are made in them than when they are made concerning the city." He said that speeches of this sort induced the young to have contempt for the established regime and made them violent. [171]

## ARISTOTLE OF STAGIRA

Aristotle of Stagira (384 BCE – 322 BCE) was a student of Plato. Many scholars believe that Aristotle had the greatest influence on future thinkers of all the ancient Greek philosophers. Aristotle's place in the history of early religious skepticism is primarily due to his enormous contributions in the establishment of the sciences. For instance, Aristotle is normally credited with developing the scientific method of inquiry which relies on careful observations of the natural world.

> [Aristotle's] own theology is an abstract monotheism, which excludes any interference on the part of the Deity in the course of the world...In the Deity he finds the final source of the coherence, order, and movement of the world, but every individual thing is to be explained in a purely natural way.[172]

Aristotle invented systems of classification for animals and plants based on these observations. In his natural

---

[171] Xenophon. *Memorabilia* (trans. by Amy L. Bonnette) Ithaca, NY: Cornell University Press (1994), Book I, Chapter 2, Sec. 9 (p. 7).

[172] Edward Zeller. *Outlines of the History of Greek Philosophy* (trans. by Sarah Frances Alleyne and Evelyn Abbott) New York: Henry Holt and Company (1889), p. 221.

sciences, the popular gods did not have a role but were simply products of myths.

> A tradition has been handed down from the ancients and from the most primitive times, and left to later ages in the form of myth, that these substances (i.e., sky and heavenly bodies) are gods and that the divine embraces all nature. The rest consists in legendary additions intended to impress the multitude and serve the purposes of legislation and the common weal; for these gods are said to have human shape or resemble certain other beings (animals)...[173]

Aristotle elaborated on the anthropomorphic character of the popular religion by comparing the social organization of gods and man.

> Wherefore men say that the gods have a king, because they themselves either are or were in ancient times under the rule of a king. For they imagine, not only the forms of gods, but their ways of life to be like their own. [174]

Like Socrates, Aristotle was charged with impiety. And like Socrates, the motivations behind the charge were political – not religious. Aristotle had been the tutor of Alexander the Great and this association with the Macedonians was resented because they ruled all Greece. Aristotle fled to escape his enemies and died shortly afterwards. He is said to have commented on his flight that he wished to prevent another sin against philosophy.

---

[173] Aristotle *Metaphysics*. Cited in Drachmann, p. 85. Drachmann explains that, "The 'sky'-gods of Aristotle have nothing in common with the gods of popular belief, not even their names, for Aristotle never names them. And the rest, the whole crowd of Greek anthropomorphic gods, exist only in the human imagination." (p. 86)

[174] Aristotle. *Politics*, 1252b 23-26. Richard McKeon (ed.) *The Basic Works of Aristotle* New York: Random House (1941), p. 1129.

# EUHEMERUS OF MESSENE

Euhemerus of Messene 'the Atheist' (fl. c. 300 BCE) is best known as the author of a work of fiction, *Sacred History*, translated into Latin by the Roman poet Ennius, which argued that the ancient myths about the gods were distortions of actual history about mortal heroes who were later worshipped as gods.

Now, Euhemerus, who was a friend of King Cassander and was required by him to perform certain affairs of state and to make great journeys abroad, says that he traveled southward as far as the ocean; for setting sail from Arabia the Blest [present day Yemen] he voyaged through the ocean for a considerable number of days and was carried to the shore of some islands in the sea, one of which bore the name of Panchaea. On this island he saw the Panchaeans who dwell there, who excel in piety and honour the gods with the most magnificent sacrifices and with remarkable votive offerings of silver and of gold. The island is sacred to the gods, and there are a number of other objects on it which are admired both for their antiquity and for the great skill of their workmanship, regarding which severally we have written in the preceeding books. There is also on the island, situated upon an exceedingly high hill, a sanctuary of Zeus Triphylius, which was established by him during the time when he was king of all the inhabited world and was still in the company of men. And in this temple there is a stele of gold on which is inscribed in summary,in the writing employed by the Panchaean, the deeds of Uranus and Cronus and Zeus. Euhemerus goes on to say that Uranus was the first to be king, that he was an honourable man and beneficent, who was versed in the movement of the stars, and that he was also the first to honour the gods of the heavens with sacrifices, whence he was called Uranus or "Heaven". There were born to him by his wife Hestia two sons, Titan and Cronus, and two daughters, Rhea and Demeter. Cronus became king after Uranus, and marrying Rhea he begat Zeus and Hera and Poseidon. And Zeus, on succeeding to the kingship, married Hera and Demeter and Themis and by them he had children, the Curetes by the first named, Persephone by the second, and Athena by the third. And going to Babylon he was entertained by Belus, and after that he went to the island of Panchaea, which lies in the ocean, and here he set up an alter to Uranus, the founder of his family. From there he passed through Syria and came to Casius, who was the ruler of Syria at that time, and who gave his name to Mt. Casius. And coming to Cilicia he conquered in battle Cilix, the governor of the region, and he visited very many other nations, all of which paid honour to him and publicly proclaimed him a god. [175]

In some respects, the views of Euhemerus comprised perhaps the most devastating opinion on the origins of ancient popular religion because he represented the Olympian gods as men.

The allegorical and physical explanations left…some room for the gods, but the historical explanation, which made them men who by might and cunning had forced their way to divine worship, like the Hellenistic rulers, deprived them altogether of their divine character. [176]

Euhemerus, despite the popularity of his *Sacred History*, was considered by many to be an atheist.

Whilst Stoicism got rid of the difficulty that discreditable stories were told of the gods by saying that the gods were gods but the stories, if literally understood, were not true, Euhemerism largely accepted the

---

[175] *Diodorus of Sicily* (trans. by C. H. Oldfather) London: William Heinemann Ltd. (1970), Book VI, I. 4-10 (pp. 333-335) Loeb Classical Library. Note 1 (p. 333) states that King Cassander was the first king of Macedonia after the death of Alexander the Great (301-0297 BCE). According to Euhemerus, Panchaea was an island in the ocean off the coast of present day Yemen. His accounts have been dismissed as mythical.
Sextus Empiricus explained how Euhemerus argued that heroes became gods. See: Sextus Empiricus. *Against the Physicists*, Book I, 17 (trans. by R. G. Bury) Cambridge, MA: Harvard University Press (1968), p. 11.
And Euhemerus, nick-named 'the Atheist' says – "When the life of mankind was without order, those who so far excelled the rest in strength and intelligence that all men lived subservient to their commands, being intent to gain for themselves more admiration and veneration, invented for themselves a kind of superhuman and divine authority, and in consequence were by the populace accounted Gods."

[176] Martin P. Nilsson. *A History of Greek Religion* (trans. by F. J. Fielden) 2nd ed., New York: W. W. Norton & Company, Inc. (1964), pp. 280-281.

stories as true, but said that the persons who acted and suffered in that way were not gods...The book was an immense literary success. Some people denounced it as impious: some accused Euhemerus of trying to perpetrate a fraud on the public; some, like poor Diodorus Siculus, two centuries and a half later, and certain of the Christian fathers, took it all for gospel truth. Probably Euhemerus no more intended his travels to be taken seriously than Voltaire did the story of *Candide*, though we may believe that he was as serious as Voltaire was in the views which he wanted his romance to convey.[177]

However, like many philosophers of his times, Euhemerus apparently believed that the heavenly bodies were gods even though he rejected the mythological gods.[178]

[Euhemerus], we know, regarded the divinities of fable as superior mortals, to whom after death the gratitude nor admiration of the multitude had accorded worship. In thus attributing to the Olympians of old no longer merely human form but also human nature, it left to the eternal and incorruptible stars alone the dignity of original gods, and exalted them in proportion as it lowered their rivals of bygone days.[179]

One modern historian explains the gradual disbelief in the fantastic creatures and happenings of Greek mythology by reference to human experience.

By means of what we would call the doctrine of present things. The past resembles the present, or, in other words, the marvelous does not exist. Now today, men with bull's heads are rarely seen, and kings do exist; therefore the Minotaur never existed, and Theseus was simply a king.[180]

---

[177] Edwyn Bevan. *Later Greek Religion* London: J. M. Dent & Sons Ltd. (1927), pp. xx-xxi.

[178] A. B. Drachmann. *Atheism in Pagan Antiquity* Chicago, IL: Ares Publishers, Inc. (1977)[reprint of the original edition published in 1922], p. 111.

[179] Franz Cumont. *Astrology and Religion Among the Greeks and Romans* New York: G. P. Putnam's Sons (1912), p. 55.

[180] Paul Veyne. *Did the Greeks Believe in their Myths?: An Essay on the Constitutive Imagination* (trans. by Paula Wissing) Chicago: The University of Chicago Press (1988), p. 14. Veyne is referring to the myth where an Athenian hero named Theseus killed the Minotaur in

## EPICURUS OF SAMOS

Epicurus of Samos (341 BCE – 271 BCE) was famous for founding a school of philosophy in Athens. Few of his writings have survived but most of the principal ideas are contained in *De Rerum Natura (On the Nature of Things)* by the Roman poet Lucretius. In Book V, Lucretius describes the gods as perfect beings that are disinterested in the affairs of men. The gods did not create the earth because it was too imperfect.

For though I were ignorant of the basic stuff,
still, just from heaven's behavior I would dare
affirm, and assert on many other grounds,
that gods most certainly never made the world
for you and me: it stands too full of flaws.
To begin: of all that heaven's vast sweep protects,
mountain, forest, and beast have rudely seized
a share; part, rocks and endless swamp possess,
and oceans, that hold our countries widely sundered.
Two-thirds besides are stolen from us mortals
by scorching heat and everlasting cold.
There's good land left, but nature's native force
would choke it with weeds if man did not fight back
for very life grown strong to heave the hoe
and curse and push the plow and slash the earth.
The clod is fertile, but the share must turn it;
our hand must harass and goad the soil to life;
on its own, it can't rise up to clear, bright air.
And sometimes, even when we've toiled and strained
and the world is all in leaf and full of flowers,
the sun in heaven burns it with fires too high,
or sudden rains destroy it, or icy frosts,
or winds with a whistle and wail whirl all away.
Besides, those terrible animals, wild and deadly
to man by land and sea: why now must nature
feed and increase them? Why do the seasons bring
disease? Why must untimely death roam free? [181]

---

the Labyrinth.

[181] Lucretius *The Nature of Things* (trans. by Frank O. Copley) New York: W. W. Norton & Co. (1977),

Epicurus believed that the gods were perfect and didn't want praise from men. He taught that men should not fear the gods and that the highest good was true knowledge.

Piety isn't to come in humble access,
on view and vested, daily to stone and altar,
to fall prostrate on earth and spread the hands
in hallowed halls, to flood the temple floors
with animal blood, and string prayer to prayer,
but the power to view a world without dismay. [182]

Like other ancient Greek philosophers, Epicurus speculated on the origins of evil and what that implied about the nature of the gods.

God, [Epicurus] says, either wishes to take away evils, and is unable; or He is able, and is unwilling; or He is neither willing nor able, or He is both willing and able. If He is willing and is unable, He is feeble, which is not in accordance with the character of God; if He is able and unwilling, He is envious, which is equally at variance with God; if He is neither willing nor able, He is both envious and feeble, and therefore not God; if He is both willing and able, which alone is suitable to God, from what source then are evils? Or why does He not remove them? [183]

---

Book V, 195-221, pp. 117-118. Also, see Lucretius. *De Rerum Natura* (trans. by W. H. D..Rouse) London: William Heinemann Ltd. (1975), Book 5, 200-217 (p. 395).

[182] Ibid., 1198-1203, p. 140.

[183] William Fletcher (trans.). *The Works of Lactantius* Vol. II Edinburgh: T&T Clark (1909)
[Ante-Nicene Christian Library: Translations of the Writings of the Fathers Down to A.D. 325
Vol. XXII], p.28. Lactantius (a famous Christian apologist) answered Epicurus by arguing that evil was necessary so man could know and choose good: "[God] set before [man] both good and evil things, inasmuch as He gave to him wisdom, the whole nature of which is employed in discerning things evil and good: for no one can choose better things and know what is good, unless he at the same time knows to reject and avoid the things which are evil" (p. 27) Therefore, unless we first know evil, we shall be unable to know good. But Epicurus did not see this, nor did any other, that if evils

The Epicurean gods, who neither created the world nor contravened the laws of nature in response to prayers from men, were very different from the traditional gods of Greek mythology. The gods had no role – which put their very existence in question. As a result, many believed that Epicurus was an atheist. For example, the Roman writer Cicero, in *The Nature of the Gods* cited Posidonius as follows:

> Posidonius argued in the fifth book of his work on the nature of the gods, that Epicurus did not believe the gods existed at all and that what he said about them was said merely to avoid the odium of atheism. [184]

Lucian, the satirist, wrote a story about the Epicurean view of the nature of the gods with predictably humorous results. In *Zeus Rants*, Zeus overhears Damis (an Epicurean) and Timocles (a Stoic) debate about the gods and then calls a meeting of the gods to discuss it.

> **Zeus**. Their whole discussion was about us. That confounded Damis asserted that we do not exercise any providence on behalf of men and do not oversee what goes on among them, saying nothing less than that we do not exist at all (for that is of course what his argument implied), and there were some who applauded him...That is why I called you together, gods, and it is no trivial reason if you consider that all our honour and glory and revenue comes from men, and if they are convinced either that there are no gods at all or that if there are they have no thought of men, we shall be without sacrifices, without presents and without honours on earth and shall sit idle in Heaven in the grip of famine, choused out of our old-time feasts and

---

are taken away, wisdom is in like manner taken away; and that no traces of virtue remain in man, the nature of which consists in enduring and overcoming the bitterness of evils." (p.28) Early Christians considered Epicurus and his followers to be atheists. In Dante's *Inferno*, Epicurus is in the 6th circle of hell (cantos 9-10).

184 Marcus Tullius Cicero. *The Nature of the Gods* (trans. by Horace C. P. McGregor) London: Penguin Books, Book I, 123, p. 120. See also, H. Rackham (trans.). *Cicero: De Natura Deorum, Academica* Cambridge, MA: Harvard University Press (2005), p. 119.

celebrations and games and sacrifices and vigils and processions. Such being the issue, I say that all must try to think out something to save the situation for us, so that Timocles will win and be thought to have the truth on his side of the argument…

**Momus** (another god). …I vow by Themis that it is not right to be angry either at Epicurus or at his associates and successors in doctrine if they could have formed such an idea of us. Why, what could one expect them to think when they see so much confusion in life, and see that the good men among them are neglected and waste away in poverty and illness and bondage while scoundrelly, pestilential fellows are highly honoured and have enormous wealth and lord it over their betters, and that temple-robbers are not punished but escape, while men who are guiltless of all wrong-doing sometimes die by the cross or the scourge?...And when the reciters tell them that we fall in love and get wounded and are thrown into chains and become slaves and quarrel among ourselves and have a thousand cares, and all this in spite of our claim to be blissful and deathless, are they not justified in laughing at us and holding us in no esteem?[185]

## Epicurus rejected the charge of atheism but acknowledged that his beliefs about the gods was different than those found in popular religion.

For verily there are gods and the knowledge of them is manifest; but they are not such as the multitude believe, seeing that men do not steadfastly maintain the notions they form respecting them. Not the man who denies the gods worshipped by the multitude, but he who affirms of the gods what the multitude believes about them is truly impious. For the utterances of the multitude about the gods are not true preconceptions but false assumptions; hence it is that the greatest evils happen to the wicked and the greatest blessings happen to the good from the hand of the gods, seeing that they are always favourable to their own good qualities and take pleasure in men like unto themselves, but reject as alien whatever is not of their kind. [186]

---

[185] A. M. Harmon *Lucian* 8 vols. London: William Heinemann Ltd. (1953), Vol. II "Zeus Rants", pp. 115-119 [Loeb Classical Library]

[186] Epicurus. "Letter to Menoeceus", 123-124 in Diogenes Laertius. *Lives of Eminent Philosophers* (trans. by R. D. Hicks) London: William Heinemann Ltd. (1965), pp. 649-651. See also, Cyril Bailey (trans.) *Epicurus: The Extant Remains* Oxford: Clarendon Press (1926), pp. 83,85. "And the impious man is not he who denies the gods of the many, but he who attaches to the gods the beliefs of the

Cicero criticized the atheism of Epicurus in his *De Natura Deorum* as follows:

> Well then, let us deny the existence of the sun, moon and planets, inasmuch as nothing can exist save that which we have touched or seen. And what of god himself? You have never seen him, have you? Why then do you believe in his existence? On this principle we must sweep aside everything unusual of which history or science informs us. The next thing would be for inland races to refuse to believe in the existence of the sea.[187]

Epicurean treatment of the nature of the gods was to become the foundation of 'deism' during the middle centuries of the last millennia.[188]

## OTHER RELIGIOUS UNBELIEF

In ancient Greece, there were several other notable philosophers, sophists and poets that doubted popular religious myths. Some of the most famous included Theodorus of Cyrene (5th c. BCE), Stilpo of Megara (4th c. BCE), and Strato of Lampsacos (3rd c. BCE). Sometimes, skeptics were accused of impiety, tried, then exiled. Little is known about their beliefs because nothing but a few fragments have survived – often only a brief mention in a book. Clearly, given that religious unbelief flourished in different places and at different times, it must have represented a significant number of educated persons who rejected the mythology of the popular religions. And yet, this would not have been

---

many."

[187] Cicero. *De Natura Deorum*, *Academica* (trans. by H. Rackham) Cambridge, MA: Harvard University Press (2000) I, 88 (pp. 85-87).

[188] Some famous deists include Voltaire and David Hume. Many of the founders of the U.S. were deists including Thomas Paine, Benjamin Franklin, Ethan Allen, and Thomas Jefferson.

thinkable before the invention of the sciences by the Greeks. Prior to that time, there was no other way to view the world than to accept supernatural causes and miraculous events.

# CHAPTER 4
# SOURCES OF CONFLICT WITH RELIGION IN ANCIENT GREECE

If ignorance of nature gave birth to gods, knowledge of nature is made for their destruction.
- Percy Bysshe Shelley *The Necessity of Atheism* (1811)

The conflict between science and religion was not the only source of conflict with religion in ancient Greece. Other sources of conflict included: moral criticism by pious believers, the introduction of foreign ideas through trade, the establishment of philosophic schools, changes in society due to the transformation from agrarian life to urban life, freedom of thought admitted by new democratic governments, and the evolution from city-states to nation-states though colonization and war. The beginning of the conflict between science and religion was nearly coincidental with the first moral criticism of religion, both becoming manifest around 600 BCE.

## GREEK RELIGION

To estimate the impact of sources of conflict with religion it is necessary to begin with a brief examination of religious practice in ancient Greece. The character of religion in ancient Greece was very different than others before or since. Historian Hugh Lloyd-Jones neatly summarized the main distinguishing features as follows:

Three important differences mark off the early religion of the Greeks from those of the surrounding peoples, and also from those that are familiar to us in modern times. First, it is neither monotheism nor, in the strict sense, polytheism, but something between the two; it has many gods, but from the earliest times of which we have knowledge one of these is preponderant over the rest. Secondly, it is not anthropocentric; men are the creation only of a minor god, and have only a minor status in the universe and a minor place in the attention of the gods. Thirdly, these gods are not transcendent but immanent; they do not interfere from the outside with the course of nature, but govern the inanimate world through natural processes and the animate through human passions. [189]

The historian Martin Nilsson offers the following description of a temple in ancient Greece and contrasts it with a church in the modern era.

A Greek temple, unlike a Christian church, was not a building in which the pious gathered together to offer worship; it was the god's dwelling place in which his image stood and his possessions were kept. When we hear of a Greek temple, we think of the great masterpieces of architecture, the Parthenon and others, but at the same time the religious value of these glorious shrines is obscured by secondary considerations. Aristotle was of the opinion that the tyrants who erected the first great temples built them to make work for the people. It is certain that these dynasts intended to display not only their reverence for the gods but their own power and glory.[190]

## THE STATE AND RELIGION

The historian Martin Nilsson neatly summarized the nature of state religion in ancient Greece by demonstrating the linkage between the piety of the citizens and the fortune of the state.

---

[189] Hugh Lloyd-Jones. *The Justice of Zeus* Berkeley, CA: University of California Press (1971), p. 160.

[190] Martin Persson Nilsson. *Greek Piety* (trans. By Herbert Jennings Rose) Oxford: Clarendon Press (1948), pp. 10-11. Nilsson comments on a well known inscription found in the temple at Delphi. Its meanings, then and now, illustrate the hazard of projecting modern perspectives into the distant past (p. 47).
Inscribed on Apollo's temple at Delphoi was the famous saying…'know thyself.' No saying has been so often repeated. For us, it signifies the need for self-knowledge; for its own time it meant, 'know that you are human and nothing more.'

> Greek religion was indissolubly connected with the community and its component parts, State, clan and family...The individual counted simply as a link in the chain of the clan, as one citizen of a State; to be cast out from the clan and the State was, next to death, the heaviest punishment which could be inflicted on anyone; such a man had neither home nor rest. The gods who protected the clan and the State protected also their members and showed kindness to them. Everyone was responsible for paying them reverence and must fulfill their demands, for an offence against them was avenged not only on the criminal but on his clan and State.[191]

Athens prosecuted impiety to prevent the gods from punishing the state and its citizens. However, the primary reasons for the trial and death of Socrates were political and not religious. Indeed, several other notable philosophers (the first scientists) and sophists (teachers) were charged with impiety or atheism in similar circumstances. For example, Anaxagoras was persecuted for his friendship with Pericles and Aristotle was threatened because he tutored Alexander of Macedon. There are few known instances where Athens charged anyone with impiety for purely religious reasons. Protagoras probably upset the public more because he was a sophist than for debating the existence of the gods. Perhaps the first clear example of an atheist in ancient Greece was Diagoras. He was reviled and sought for punishment. Later his name even became synonymous with atheism.

In Pre-Socratic times, both individuals and states sought the favor of the gods by blood sacrifices and other offerings. They also employed seers or applied to oracles to foretell the future. Eventually, the efficacy of oracles was questioned – even the great oracle of Apollo at Delphi.

---

[191] Martin Persson Nilsson. *Greek Piety* (trans. By Herbert Jennings Rose) Oxford: Clarendon Press (1948), pp. 7-8

That Delphi manifestly failed to foresee the Greek victory in the Persian Wars and all too clearly recommended surrender badly damaged its reputation, in spite of all attempts to reinterpret its pronouncements. Thereafter political decisions were increasingly taken without reference to the oracle.[192]

## The failure of the Greek states during the fourth century BCE, following the Peloponnesian War, led directly to a weakening of belief in state gods.

The state had no longer anything to offer its citizens, and they on their side had no reason to plead the cause of the state gods. With the dwindling of political power the religion attached to the state declined also.[193]

---

[192] Walter Burkert. *Greek Religion: Archaic and Classical* (trans. by John Raffan) Oxford: Basil Blackwell (1985), p. 116.
Herodotus relates how Persian king Xerxes was preparing to invade Greece and Athens sent messengers to the oracle at Delphi. The priestess Aristonice provided the following answer:
Wretches, why tarry ye thus? Nay, flee from your houses and city,
Flee to the ends of the earth from the circle embattled of Athens!...
A. G. Godley (trans.) *Herodotus* 4 vols. London: William Heinemann (1922) Vol. III, Book VII, 140, (p. 445)
The Athenians were persuaded to request a second answer from the oracle which advised evacuating Athens and meeting Xerxes in battle at Salamis.
...
*Yet shall a wood-built wall by Zeus all-seing be granted*
*Unto the Trito-born, a stronghold for thee and thy children.*
*Bide not still in thy place for the host that cometh from landward,*
*Cometh with horsemen and foot; but rather withdraw at his coming,*
*Turning thy back to the foe; thou yet shalt meet him in battle.*
*Salamis, isle divine! 'tis writ that children of women*
*Thou shalt destroy one day, in the season of seed-time or harvest.*
Themistocles convinced the Athenians that the second answer at Delphi meant they would win a victory over the Persians at sea near Salamis. Their warships would be the "wood-built wall." The Athenians agreed to build 200 warships and won a great victory at Salamis.
[193] Martin P. Nilsson. *A History of Greek Religion* (trans. By F. J. Fielden) 2nd ed. Oxford: Clarendon Press (1949), p. 277.

## CITY LIFE AND RELIGION

### Religion in ancient Greece evolved over time and in response to societal changes and political developments.

Religion is dependent on the conditions of life. When these change new needs arise and old forms wane, and popular religion undergoes corresponding changes. Such changes were effected when people crowded into the towns and began to earn their livelihood not by agriculture and stockbreeding but by industry and commerce. Changes in political life and the rise of democracy also caused certain changes in religion. We should bear in mind that in the democratic states the people formed the popular assembly to which all decisions pertained, even in religious matters. The result was that religion was secularized to a certain degree. But it was not dead. Religion tried to find new forms corresponding to the new needs and new ideas of the people. This movement was only beginning in the fifth and fourth centuries B.C. The real turning point is the age of the Sophists.[194]

### State religion eventually evolved into personal religion in ancient Greece because the Olympian gods were too remote and unresponsive to the daily cares of the individual.

The religious situation in Greece was complicated in the fifth and fourth centuries B.C., even in regard to popular religion. It was simple enough in backward districts, where the old faith survived without being disturbed and where the people kept rustic customs, celebrated the old festivals, and venerated the gods and heroes without doing much thinking about the high gods...the situation was different elsewhere, especially in the cities, where religion had to encounter political life and the new enlightenment. The people ascribed the greatness and glory of the state, its freedom and independence, to its great goods...But the cult of the great gods was too cold. These gods did not offer help in human needs and consolation to a contrite heart. The old bonds of state and family were loosened, the individual became conscious of himself...Man was no longer born to his gods as in earlier times. He wanted to find his gods for himself.[195]

---

[194] Martin P. Nilsson. *Greek Folk Religion* New York: Harper and Brothers (1961), p. 138.
[195] Ibid., p. 121.

## TRADE AND RELIGION

Whatever the reasons for the growth of skepticism, the increase in trade and the resulting exchange of ideas between cultures probably contributed. To promote trade, Greek colonies were established along the shores of the Mediterranean from modern day France to Turkey, including Egypt and north Africa.

Additionally, Greek colonies were founded on the shores of the Black Sea – especially by Miletus. The communication of ideas was made easier with the invention of writing, including mathematical notations, and by the spread of Greek as a common language. For instance, many scholars have written on the early influence of Oriental ideas on Greek thought and culture, centuries before the conquests by Alexander of Macedon, due to ancient overland trade routes connecting Asia with Europe.

Walter Burkert, a historian of ancient Greece, commented that tradition, the foundation of religion, was weakened by the introduction of foreign ideas.

> *Nomos*, meaning both custom and law, becomes a central concept of Sophistic thought. Laws are made by men and can be altered arbitrarily. And what is tradition if not the sum of such ordinances? Horizons are extended through travel and reports of travel: with growing interest men became aware of foreign peoples among whom everything is quite different, witness the ethnographic digressions in Herodotus. In this way the unquestioned assumptions of custom can easily be shaken...But it was on tradition, *nomos*, that religion primarily rested, as the Greeks knew well. Its foundations were seen to be threatened, at least in theory, as a result of the questioning of *nomos*. [196]

---

[196] Walter Burkert. *Greek Religion: Archaic and Classical* (trans. by John Raffan) Oxford: Basil Blackwell (1985), pp. 312-313.

The Greeks became aware of the differences and similarities between their gods and the gods of the other civilizations in Egypt and Babylonia – especially as trade and war brought them together. The educated classes understood that these differences proved that the myths of the gods were fictions and that perhaps even the gods themselves were inventions of man.

> The differences between the cults and gods of different states showed that religion was founded not in Nature but convention. There is, it is said later, one god according to Nature, but many according to law and custom. The logical consequence here also is that the gods are the work of man. [197]

## HISTORY AND RELIGION

Historians in ancient Greece viewed the gods and religion more objectively than the common man. For example, Herodotus, who traveled extensively, wrote:

> I do not think that any one nation knows much more about [religion] than any other. [198]

After him, the historian Thucydides was even more unwilling to attribute world events to acts of the gods.

> The strongest evidence as to the free-thinking of the Periclean Age is, however to be met within the historical writings of Thucydides. In his work on the Peloponnesion War, Thucydides completely eliminated the supernatural element; not only did he throughout ignore omens, and divinations, except in so far as they played a part as a psychological factor, but he also completely omitted any reference to the gods in his narrative. Such a procedure was at this time unprecedented, and contrasts sharply with that of his immediate forerunner Herodotus, who constantly lays stress on the intervention

---

[197] Martin P. Nilsson. *A History of Greek Religion* (trans. by F. J. Fielden) 2nd ed., New York: W. W. Norton & Company, Inc. (1964), p. 269.

[198] Herodotus. *The Histories* (trans. by Aubrey de Slincourt) New York: Penguin Books (1972), Rev Ed., Book II, p. 130.

of the gods. That is hardly conceivable except in a man who had altogether emancipated himself from the religious views of his time.[199]

For example, in his *History of the Peloponnesian War*, Thucydides has the Athenian envoys plead with the Melian leaders not to rely on religious beliefs to deliver them from certain destruction if they fail to surrender.

This fate, we beg of you, weak as you are and dependent on a single turn of the scale, do not willingly incur; nor make yourselves like the common crowd who, when it is possible still to be saved by human means, as soon as distress comes and all visible grounds of hope fail them, betake themselves to those that are invisible – to divination, oracles, and the like, which, with the hopes they inspire, bring men to ruin.[200]

Another example of the weakness of religious beliefs in Athens at the close of the fifth century BCE is found in Thucydides where he relates a funeral speech of Pericles for the first dead in the Peloponnesian War (c. 431 BCE).

It cannot possibly be due to chance that the magnificent funeral oration of Pericles, which inevitably included some concessions to convention in addition to its nobler contents, should have omitted an allusion to the figures of the popular religion. The omission was obviously intentional, and we recognize the historian's wish to characterize the free-thinking pupil of Anaxagoras in his true colours

---

[199] A. B. Drachmann. *Atheism in Pagan Antiquity* Chicago, IL: Ares Publishing Inc. (1977), pp. 28-29. See also: Will Durant. *The Life of Greece* New York: Simon and Schuster (1939) [The Story of Civilization, Part II], p. 435. Regarding Thucydides' writings Durant commented:

Here are no legends, no myths, and no miracles. He accepts the heroic tales, but tries to explain them in naturalistic terms. As for the gods, he is devastatingly silent; they have no place in his history. In him the conflict between religion and philosophy is decided; and philosophy wins.

[200] Thucydides. *History of the Peloponnesian War* Vol. III (trans. by Charles Forster Smith) London: William Heinemann Ltd. (1952), Book V, C III, p. 167.

as the champion of philosophic doubt in respect to the whole system of mythology.[201]

By the end of the Pelopponesian War, following the devastating plague at Athens, the failure of the gods to save the Athenians and the apparently random deaths of even its most pious citizens greatly undermined state religion. Thucydides attributed the general lawlessness and civil disorder that accompanied the plague to this.

No fear of gods or law of men restrained; for, on the one hand, seeing that all men were perishing alike, they judged that piety and impiety came to the same thing, and, on the other, no one expected to be called to account and pay the penalty for his misdeeds. [202]

## PHILOSOPHY AND RELIGION

New philosophies, such as Epicureanism and Stoicism, superseded popular religion based upon myths of the gods for some in ancient Greece and Rome[203] but Stoicism was eventually replaced by Christianity.

People in Hellenistic times had lost the support and comfort which a living religious faith can give against the blows of fortune and the unwelcome events of life. They were like men driven hither and thither at sea without a rudder, at the mercy of the waves. Criticism and unbelief had wrenched the tiller from the hands of the gods, but philosophy offered men a rudder and prompted them to steer their ship along a course which aimed at freeing them from the blows and arbitrariness of (Fortune)…Philosophy took the place of religion as the comforter and guide of humanity, and there is plenty of evidence that it was able to give mankind power to stand up to the storms in

---

201 Theodor Gomperz. *Greek Thinkers: A History of Ancient Philosophy* 2 vols. (trans. by Laurie Magnus) London: John Murray (1920) Vol. I, p. 514 [orig. publ. 1896].

202 Charles Forster Smith (trans.) *Thucydides* 4 vols. London: William Heinemann (1928), Vol. II, LIII (p. 353).

203 See Lucretius *The Nature of Things* which promoted Epicureanism. Cicero argued against Epicureanism in his work *The Nature of the Gods*.

life; but it was a view of life fit only for the educated, who thought seriously about existence, not for the great mass of humanity, which needed, not philosophical maxims, but a firm concrete belief.[204]

The new philosophical schools appealed most to the educated man. The common man generally remained loyal to the gods of mythology or fell back on the traditional nature religion.

The anthropomorphic gods had lost their power over the minds of the educated, who devoted themselves instead to the art of living. To pilot mankind through the surges of life was now the main task of philosophers; they became shepherds and confessors of souls. Both the Stoic self-sufficiency and the Epicurean quietism were objective methods of viewing the course of the world…[205]

---

[204] Gilbert Murray. *Four Stages of Greek Religion* New York: Columbia University Press (1912), p. 87-88. Ancient Greek history is divided into the Archaic Period (700-500 BCE), Classical Period (500-323 BCE), and the Hellenistic Period (323-31 BCE). Murray elaborates and extends this theme (p. 91).

…(R)eligious feeling never was lower than when the Hellenistic period began. It was replaced by belief in one's own power and trust in (Fortune), and, among the educated, by the philosopher's doctrine concerning life. For the common people there remained cult and the lower forms of religion, supplemented by superstition, which usually increases in a time when the old religions are falling.

[205] Martin P. Nilsson. *A History of Greek Religion* (trans. By F. J. Fielden) 2nd ed. Oxford: Clarendon Press (1949), p. 282.

Unbelief is seldom unaccompanied by some feeling of uneasiness. There is so much in heaven and earth that philosophy, and still more the plain man, has never dreamed of: doubt comes upon him whether his simple everyday materialism will see him through. The inherited ideas of something supernatural, of gods, will not let him go. In times of need and misfortune, when he is abandoned by men, he shrinks back from the empty void he has created. This fundamental religious feeling was more tenacious than popular unbelief or philosophical enlightenment. But it did not benefit the ancient gods, for they were worn out.

The Cynic school of philosophic thought asserted that the transition of Greek society from rural living to city life marked the beginning of the degradation of man.

In general, every creature is capable of living in the situation in which it is naturally placed. Otherwise, the first men could not have maintained their existence, for they lacked the use of fire just as much as dwelling-places, clothing, and artificially prepared food. Over-subtlety and the busy spirit of invention have done little to bless mankind. The greater men's efforts to obviate the hardships of life, the harder and the more toilsome has life become. And herein lies the true significance of the Prometheus-myth. The Titan was not punished because Zeus hated the human race, but because that gift of fire had sown the seeds of civilization, and therewith those of luxury and all corruption.[206]

## THE SOPHISTS AND RELIGION

The sciences revealed natural causes for phenomena that had previously been attributed to acts of the gods. However, the spread of these ideas was largely the work of sophists (teachers) who were active throughout the Greek mainland and its colonies beginning with Protagoras in the fifth century before the common era. The books of Protagoras were burned in Athens after he questioned the existence of the gods.

During the Peloponnesian war, when Athens was struggling for her existence, she was more than usually sensitive to anything which might imperil her well-being, at the hands of the gods no less than other ways. The greatest danger came from the mighty intellectual awakening, the period of enlightenment, brought about by the work of the sophists. They did not, like the philosophers, appeal to a limited circle nor work, as they did, principally through their writings, but came forward as teachers of the young, and the young flocked to their lectures and discussions...All the questions of the day were drawn into the discussions, not excepting the natural philosopher's explanations of the phenomena of the universe. What

---

[206] Theodor Gomperz. *Greek Thinkers: A History of Ancient Philosophy* 2 vols. (trans. by G. G. Berry) New York: Charles Scribner's Sons (1905) Vol. II, p. 145.

had formerly been as esoteric science now formed a part of general culture. Thus the explanations of the natural philosophers stood out in opposition to the religious ideas, and the conflict became real. [207]

Socrates, a contemporary of Protagoras, was accused of being a sophist because his enemies sought to turn public opinion against him as they believed that he taught ideas subversive of the democracy in Athens. However, Socrates was opposed to the scientific study of heavenly bodies except for practical applications. In his *Memorabilia*, Xenophon relates that Socrates,

> ...strongly deprecated studying astronomy so far as to include the knowledge of bodies revolving in different courses, and of planets and comets, and wearing oneself out with the calculation of their distance from the earth, their periods of revolution and the causes of these. Of such researches, again he said that he could not see what useful purpose they served...
>
> In general, with regard to the phenomena of the heavens, he deprecated curiosity to learn how the deity contrives them: he held that their secrets could not be discovered by man, and believed that any attempt to search out what the gods had not chosen to reveal must be displeasing to them. [208]

Despite his conventional piety, Socrates is generally considered one of the earliest, and most famous, casualties in the conflict between science and religion.

The prosecution of Socrates was one of the first acts of the restored democracy after the expulsion of the thirty tyrants. After him various philosophers were accused of atheism. The accusation of Aristotle had a political motive in the hatred of Alexander, but no such motive was present in the cases of Stilpo the Megarian, Bion the Cynic, or Theodoros the Cyrenaean. In the following century, the third, the religious prosecutions ceased, although such scepticism as that of Carneades might just as well have led to prosecution as the criticisms

---

[207] Martin P. Nilsson. *A History of Greek Religion* (trans. by F. J. Fielden) 2nd ed. Oxford: Clarendon Press (1949), pp. 265-266.

[208] E. C. Marchant (trans.) *Xenophon: Memorabilia and Oeconomicus* London: William Heinemann Ltd. (1938), IV, vii 5-6, p. 349.

of Protagoras. The religious indifference which was the result of the undermining work of the critics had seized upon the whole people, even the masses.[209]

## LITERATURE AND RELIGION

The conflict between the poets and religion is amply documented in the tragic themes explored by Euripides and in the absurd comedy of Aristophanes. In the plays of Euripides, the flourishing of evil was proof that the gods were indifferent or that the gods did not exist. Aristophanes made fun of the speculations of Socrates and the sophists but he also directed his humor at some religious practices such as divination and oracles. Their plays reflected the mood of the times and remained very popular for many years.

It is beyond doubt that the criticisms of myths and religious practices by the Sophists and by the dramatists such as Euripides and Aristophanes had a major influence on Greek theology and philosophy in the late fifth century and afterwards.[210]

Much of the poetry of ancient Greece explored religious themes and posed the age old questions about the meaning of life. Semonides of Samos (c. 29th Olympiad, 664-661 BCE) was typical of these early poets and left a famous fragment commenting that man's life was short and filled with evil.

Thundering Zeus, lad, hath the ends of all things there be, and doeth with them what he will. There's no mind in us men, but we live each day as it cometh like grazing cattle, knowing no whit how God shall end it...There's none alive but thinketh he will come home winged with wealth and good things next year; yet one of us ere he reach his goal is taken with unenvied Age, another's mind is wasted by

---

[209] Martin P. Nilsson. *A History of Greek Religion* (trans. by F. J. Fielden) 2nd ed. Oxford: Clarendon Press (1949), p. 275.
[210] Jon D. Mikalson. *Athenian Popular Religion* Chapel Hill, NC: University of North Carolina Press (1983), p. 112.

miserable Disease, or Death sendeth him below dark Earth whelmed by War. Some die at sea when they have laden a ship with their substance, confounded by storm and the many waves of the purple brine; others tie a noose about their miserable neck and leave the sunlight of their free choice. So true it is that nothing is without ills, nay, ten thousand the Dooms of men, and their woes and sorrows past reckoning. [211]

The questioning by the poets challenged religious belief just like the theories advanced by the natural philosophers. The origin of this revolutionary thinking was in Ionia (see Chapter 1).

...(B)oth are typical products of the same general state of society, presupposing alike intellectual freedom, a high material standard of life, and abundant leisure. The spirit of their age and class was materialistic. Their interests lay in this world, they had not much faith in another...The gods had been politely shelved, and divine expectations of the origin and nature of the world were no longer tolerable. The question was raised, and raised for the first time, whether man's intellect could solve these problems without recourse to myth. [212]

## MORAL CRITICISM OF RELIGION

Nearly coincidental with what we now recognize as the beginnings of science, there arose a moral criticism of the gods of mythology as depicted in the poems of Homer and Hesiod. This moral criticism was pointed and powerful whereas science, initially, only indirectly undermined popular belief in the traditional religion.

---

[211] J. M. Edmonds (trans.). *Elegy and Iambus: Being the Remains of All the Greek Elegiac and Iambic Poets* 2 vols. London: William Heinemann Ltd. (1931), Vol. II, "Semonides", pp. 213, 215. Also, cited by W. K. C. Guthrie. *The Greeks and Their Gods* Boston: Beacon Press (1954), p. 131.

[212] W. K. C. Guthrie. *The Greeks and Their Gods* Boston: Beacon Press (1954), p. 132.

Ironically, some of the earliest criticism of popular religion in ancient Greece was by philosophers who themselves held very strong traditional religious beliefs. For example, Xenophanes criticized anthropomorphism in the mythology and he disbelieved in divination (e.g., oracles). Similarly, Heraclitus objected to idolatry and blood sacrifices that were often depicted in the myths. As a result, these thinkers argued that the popular religious mythology itself was impious and immoral. They were not regarded as atheists, even in their own times, but rather moralists.

Xenophanes and others challenged the morality of the gods and their anthropomorphic character as portrayed by the poets. Xenophanes objected that the human traits ascribed to the gods included the worst features of man-lust, jealousy, violence and deceit. To Xenophanes, it was obvious that men imagined the gods to have human appearance and they naturally added human emotions and motives. However, Xenophanes' rejection of these inventions was testimony of his strong religious beliefs and not, as some have supposed, evidence of his skepticism. He did not challenge the existence of the gods but only their characterization by Homer and Hesiod. Similarly, more than a century later, Euripides was said to have written, "If the gods do anything base, they are not gods."[213]

---

[213] Quoted in Martin P. Nilsson. *A History of Greek Religion* (trans. by F. J. Fielden) 2nd cd., New York: W. W. Norton & Company, Inc. (1964), p. 274. Nilsson commented, "With this saying, the gods of Greek mythology were doomed."
See also, Arthur S. Way (trans.) *Euripides* Vol. IV 4 vols, London: William Heineman (1912), p. 47 "Ion" (l. 449-451).

*Unjust it were*
*To call men vile, if we but imitate*
*What Gods deem good: - they are vile*
*Who teach us this.*

The moral criticism of the mythological gods advanced by Xenophanes and Heraclitus was not taken up by others until the time of Periclean Athens.

> The scepticism regarding the gods expressed by Xenophanes seems not to be echoed by any of his contemporaries, nor by any writer before the middle of the fifth century. But from that time on scepticism and atheism became forces to be reckoned with. [214]

The anthropomorphic gods of Homer and Hesiod were distant and didn't have the immediacy of the nature gods. The weakening of the influence of these gods was attributed, in part, with the flourishing of science.

> The Olympian God shed his functions, and so is cut loose from his anchorage in the life of earth. In proportion as he becomes anthropomorphic, he comes to be less and less in touch with his natural religion. The formation of a pan-hellenic religion meant that the Olympians left their provinces to go and form a polity of their own on the top of a mountain. Finally, they leave the earthly Olympus, and vanish into the sky. Nature, depeopled of Gods is left free for science. [215]

Further proof of Xenophanes' strong religious beliefs can be found in his argument that it was impious to hold that the gods were born or died because it implied that the gods did not always exist. [216] The fact that he made

---

Finally, see Lewis Richard Farnell. *Outline-History of Greek Religion* Chicago: Ares Publishing (1974),
p. 114.
[The plays of Euripides] are directed mainly and most forcibly against the stories of divine vindictiveness and divine licentiousness. [Euripides] is evidently touched with the new idea that vengeance is alien to the perfect nature of God...

[214] Hugh Lloyd-Jones. *The Justice of Zeus* Berkeley, CA: University of California Press (1971), p. 129.

[215] F. M. Cornford. *From Religion to Philosophy: A Study in the Origins of Western Speculation* New York: Harper Torchbooks (1957), p. 115.

[216] "It was a saying of Xenophanes that to assert that the gods had

this statement is evidence that the Greeks still believed in the mythical stories concerning the origins of the gods. More importantly, if the gods didn't always exist, it meant that they didn't create the universe but simply lived in it like mortal men.

The result of moral criticism was that the powers and stature of the gods were diminished. This tended to encourage further speculation about the nature of the gods and even provoked questions about their existence. The well known historian J.B. Bury speculated on the origins of religious skepticism in ancient Greece:

> Now let us suppose that before the end of the sixth century a thoughtful man began to reflect on the past fortunes of the Greeks. He would be struck by the fact that the character of their history had completely changed. The age of the heroes, as described in the epics, was marked by divine interventions, frequent intercourse between gods and men, startling metamorphoses, and all kinds of miracles. How was it that the character of human experience had changed and that such marvels had ceased to happen? It was inevitable that the question should be asked: can we believe the epic poets and take all they tell us for literal fact? [217]

## THE PROBLEM OF EVIL

Another strong moral argument related to the nature of the gods was the problem of evil. The fact that evil existed raised questions about the gods regarding their concern for man and their power to overcome evil. Ultimately, for some, these arguments led to a practical atheism where nature became god.[218] Pliny the Elder, the

---

birth is as impious as to say that they die; the consequence of both is that there is a time when gods do not exist." Aristotle. *Rhetoric* 1399b 5-8 in Richard McKeon (ed.) *The Basic Works of Aristotle* New York: Random House (1941), p. 1425.

[217] J. B. Bury. *The Ancient Greek Historians* New York: The Macmillan Co. (1909), p. 10.

[218] The Victorian poet Alfred Tennyson expressed his questions about

famous Roman encyclopedist, elaborated on this theme in his *Natural History* where he concluded that,

> But the chief consolations for nature's imperfection in the case of man [re: man's wickedness] are that not even for God are all things possible – for he cannot, even if he wishes, commit suicide, the supreme boon that he has bestowed on man among all the penalties of life, nor bestow eternity on mortals or recall the deceased, nor cause a man that has lived not to have lived or one that has held high office not to have held it – and that he has no power over what is past save to forget it, and (to link our fellowship with God by means of frivolous arguments as well) that he cannot cause twice ten not to be twenty or do many things on similar lines: which facts unquestionably demonstrate the power of nature, and prove that it is this that we mean by the word 'God.'[219]

Early myths explained the existence of evil as punishment of man for seeking knowledge held by the gods. For example, Prometheus gave fire to man and, as punishment, Zeus caused Pandora to open a jar which allowed evil to escape into the world.

Cicero, the prominent Roman author and statesman, wrote about the problem of the existence of evil in *The Nature of the Gods*. In the dialogues, the character Cotta

---

the nature of god because of the existence of natural evil in his famous poem "In Memoriam" (1849), Section 56, Lines 13-16:

*Who trusted God was love indeed*
*And love Creation's final law –*
*Tho' Nature, red in tooth and claw*
*With ravine, shrieked against his creed –*

Robert H. Ross. *Alfred,Lord Tennyson: In Memoriam* New York: W. W. Norton & Company Inc. (1973), pp. 35-36. A similar sentiment was expressed by the poet William Blake (1757-1827) in his popular short poem "The Tiger."

*Tiger, tiger, burning bright*
*In the forests of the night,*
*What immortal hand or eye*
*Could frame thy fearful symmetry?*

[219] Pliny. *Natural History* Vol. I London: William Heinemann Ltd. (1938), Book II, v. 27 (p.187).

points out that the gods must have allowed evil to exist but disagreed with the explanation that it serves as punishment for man's wrongdoing since the gods could have created man so that he could not do wrong.

> You argue that man is to blame for all this and not the gods. As well might a doctor blame the gravity of a disease or a pilot the violence of the storm. Coming from a mere man such an excuse would be ridiculous. "Who would have any need of your services," one would reply, "if there were no diseases and no storms?" A god one could rebut more forcibly still. "You gods say that the fault lies in the vices of mankind. But you could have endowed men with reason in a form which would exclude all vice and crime." How then did the gods come to make such a mistake? [220]

For Epicurus, the existence of evil in the world was key to shaping his teachings about the nature of the gods. He believed that the gods must be either unwilling or unable to remove evil and he concluded that the gods were unconcerned with man. His humorous comment on the efficacy of prayer also makes this point.

---

[220] Cicero. *The Nature of the Gods* (trans. by Horace C. P. McGregor) New York: Penguin Books (1972), Book III, 76. See also, H. Rackham (trans.) *Cicero: De Natura Deorum, Academica* Cambridge, MA: Harvard University Press (2005), pp. 361, 363.
This is not an argument against free will. Charles Darwin made similar remarks in his *Autobiography*. "That there is much suffering in the world no one disputes. Some have attempted to explain this in reference to man by imagining that it serves for his moral improvement. But the number of men in the world is as nothing compared with that of all other sentient beings, and these often suffer greatly without any moral improvement. A being so powerful and so full of knowledge as a God who could create the universe, is to our finite minds omnipotent and omniscient, and it revolts our understanding to suppose that his benevolence is not unbounded, for what advantage can there be in the sufferings of millions of the lower animals throughout almost endless time? This very old argument from the existence of suffering against the existence of an intelligent first cause seems to me a strong one..." Nora Barlow (ed.)*The Autobiography of Charles Darwin 1809-1882* New York: W. W. Norton & Company (1969), p.90.

If god listened to the prayers of men, all men would quickly have perished: for they are for ever praying for evil against one another. [221]

A more troublesome aspect of evil for ancient philosophers was not simply that evil existed but that it prospered. Diogenes of Sinope pointed out that often people who committed evil acts lived long lives and prospered. Euripides, in his play *Bellerophon*, explored this theme by commenting on the success of evil tyrants in defeating cities where the citizens were pious but had smaller armies. Diagoras of Melos, considered by many to have crossed from mere religious skepticism to outright disbelief in the existence of the gods, was said to have asserted that the character of a person did not affect his fortune. The poet Theognis of Megara (6th c. BCE) wrote elegies that address this aspect of evil and, in particular, the injustice where the punishment for wrongdoing is delayed by generations then visited upon the innocent.

Dear father Zeus, I'm willing for the gods
(If they insist) to let a villain love
His outrage, but I wish that, after that,
The criminal, who acts intentionally
And disregards the gods, would pay, himself,
The penalty for crime – not that the sins
Of fathers should bring sorrow to their sons
In later days. The evil father's sons
Who practice justice, son of Kronos, they
Who fear your anger, loving from the first
The Right among their fellow-citizens,
Should not be punished for their father's crimes.
I wish the gods agreed. But as things are,
Bad men escape, and others bear the brunt.[222]

---

[221] Cyril Bailey (trans.) *Epicurus: the Extant Remains* Oxford: Clarendon Press (1926), p.135 [Fragment 58]

[222] Dorothea Wender (trans.). *Hesiod and Theognis* London: Penguin Books (1987), p. 122, l. 731-42. It is interesting that

Closely related to the argument that evil prospers was the question of whether evil went unpunished. Sextus Empiricus wrote a famous passage in his book *Against the Physicists* where he argued that it was not possible to punish the dead because they are no longer conscious. Nevertheless, it was generally believed by the Greeks that evil was punished in Hades in the afterlife. Of course, this thinking also persisted in western religions that eventually replaced pagan beliefs.

In Plutarch's *On the Delay of the Deity in the Punishment of the Wicked*, the character of Epicurus speaks to the group and his arguments are summarized by other participants in the dialogue. For example, Patrocleas, Plutarch's son-in-law, comments on aspects of the discourse as follows.

> (Epicurus) asserts, that the long delayed punishments of the guilty argues an unbecoming indifference on the part of the Deity to the deserts of crime; that it fails to restrain the commission of it, as a more speedy retribution would do; that it is often entirely useless as a reparation to those who have suffered from injustice, since they are generally dead long before their wrongs are vindicated; and finally, that it emboldens the wicked in their course by leading them to think, that because they are spared for a time, they may therefore trample on the laws of God with permanent impunity.[223]

---

Theognis calls Zeus the son of Kronos when appealing to him about the injustice of punishing sons for their father's crimes. Of course, Greek myth represents Kronos as devouring his children , except Zeus who escaped, to present a challenge to his power.

See also, Douglas E. Gerber (trans.). *Greek Elegiac Poetry: From the Seventh to the Fifth Centuries B.C.* Cambridge, MA: Harvard University Press (1999), lines 731-52 (pp. 281-283).

[223] Plutarch. *On the Delay of the Deity in the Punishment of the Wicked* (trans. By H. B. Hackett) New York: Mark H. Newman (1844), p. 62.

See also, Phillip H. De Lacy and Benedict Einarson. *Plutarch's Moralia* 15 vols. London: William Heinemann Ltd. (1959) Vol. VII, "On the Delays of the Divine Vengeance."

Another participant in the dialogue, Olympicus, elaborates on the discourse of Epicurus with his own comments.

Olympicus ...adds to these objections still others. He urges especially, that such delay in the punishment of the wicked must have a tendency to destroy the belief of men in the reality of a divine Providence itself; and further, that where the act and its consequences follow each other at such a distance, no useful disciplinary purpose can possibly be accomplished, as regards the reformation of the offender himself. Punishment which is delayed a long time after the commission of the crime for which it is inflicted, instead of being referred to its proper source, is more likely to be regarded as a calamity or accident, and thus, though it may serve to annoy and distress the guilty person, must fail entirely to lead him to profitable reflections on his conduct and the cause of his suffering.[224]

Plutarch, speaking for himself, seeks to vindicate God against the accusations of the character Epicurus by enumerating several possible explanations for the delay in punishing the wicked. These include: setting an example of right conduct with respect to those who harm us, giving the wicked time to repent, allowing the wicked to be used to reform others, and making the punishment more conspicuous. Plutarch's brother Timon protests that it is improper to punish children for the evil of their parents. Plutarch answers that this punishment prevents similar evil by the offspring since children inherit qualities from their parents. Further, parents are prevented from committing acts of evil since they don't wish to see their offspring punished.[225]

Much later, Christian apologist St. Augustine answered the same question in his *Treatise on the City of God.*

Some one may ask, 'Why, then, was that divine mercy extended to the ungodly and ungrateful? ' Why, but because it is from Him Who

---

[224] Ibid., pp. 62-63.
[225] Ibid., Plutarch – pp. 64-67, Timon – p. 68.

maketh His 'sun to rise on the evil and on the good, and sendeth rain on the just and on the unjust'?...It is a divine patience that invites the wicked to repentance, while it is divine penance that educates the good in patience...Yet all the time God's working can be plainly seen in the present distribution of good and evil; for, if every sin were now clearly punished, nothing would seem to be reserved for the last judgement; while, on the other hand, if no sin were manifestly punished here, the providence of God might be denied. In like manner, with regard to the good things of this world; if God did not respond to some requests with abundant kindness, men might say that He had not such things at His disposal; while if every such petition were answered, one might suppose that it was only right to serve Him for these worldly things...[226]

Hesiod, in *Works and Days*, described how the gods keep watch on evil deeds:

You princes, mark well this punishment you also; for the deathless gods are near among men and mark all those who oppress their fellows with crooked judgements, and reck not the anger of the gods. For upon the bounteous earth Zeus has thrice ten thousand spirits, watchers of mortal men, and these keep watch on judgements and deeds of wrong as they roam, clothed in mist, all over the earth.[227]

Still another aspect of evil that preoccupied early Greek philosophers was the fact that evil in nature harmed both good and evil people equally. An anecdote about Diagoras of Melos has him commenting on how storms at sea drown everyone. Thucydides, in his history on the Peloponnesian War, observed that the plague in Athens killed indiscriminately.[228]

---

[226] F. R. Montgomery Hitchcock (trans.). *St. Augustine's Treatise on the City of God* New York: The Macmillan Company (1922), Book I "The Church's Answer to Paganism". Ch. VIII, pp. 23-24.
[227] Hesiod. *Works and Days*, lines 248-255 in Hugh G. Evelyn-White (trans.) *Hesiod: The Homeric Hymns and Homerica* London: William Heinemann Ltd (1982), p. 21.
[228] Thucydides. *The History of the Peloponnesian Wars*, II, 53.

## FATE AND RELIGION

Beginning in the 4th century BCE, a growing number of the uneducated masses rejected belief in the Olympian gods and embraced, instead, a belief in fortune or destiny. This was probably inevitable given the problem reconciling evil with religion.

> It is worth remembering that the best seed-ground for superstition is a society in which the fortunes of men seem to bear practically no relation to their merits and efforts...And in such a society people tend to lay stress on the reasonable or visible chains of causation. But in a country suffering from earthquakes or pestilence, in a court governed by the whim of a despot, in a district which is habitually the seat of a war between alien armies, the ordinary virtues of diligence, honesty and kindliness seem to be of little avail. The only way to escape destruction is to win the favour of the prevailing powers, take the side of the strongest invader, flatter the despot, placate the Fate or Fortune or angry god that is sending the earthquake or the pestilence. The Hellenistic period pretty certainly falls in some degree under all of these categories. And one result is the sudden and enormous spread of the worship of Fortune.[229]

Belief in fate led the Greeks to a fascination with astrology that was to persist with the Romans and even into modern times. Plato was responsible, in large part, for persuading the Greeks that heavenly bodies were, in fact, visible gods. It was inevitable that astrology would enjoy enormous popularity as a result. Historian Walter Burkert believed that astrology also found some

---

[229] Gilbert Murray. *Four Stages of Greek Religion* New York: Columbia University Press (1912), pp. 112-113. Murray succinctly states that belief in Fate allows for the practice of divination (pp. 125-126):

If a prophet can fortell that such and such an event will happen, then it is obviously fated to happen. Foreknowledge implies predestination.

Of course, the implication is that appeals to the gods are fruitless since events are predestined – an idea contrary to the concept of free will in popular religious practice.

scientific support due to early astronomical discoveries that found order in the universe. According to Burkert, after Plato the mythical gods were replaced by visible gods.

Cosmos religion and star religion are henceforth, especially in the Hellenistic Age, the dominant form of enlightened piety…Certainly it was always possible to identify the mythical with the visible gods through the method of allegory. The Stoics in particular were responsible for carrying this out in a detailed way: many of their equations became the common property of all educated people down to the age of the Baroque: Zeus is the sky, Apollo the sun, Artemis the moon, Demeter the earth. The planets, which are less obvious to the layman, failed to attain a similar popularity; yet astrology, based on the calculation of their periods, became from the Late Hellenistic period onwards, a dominant spiritual force as a new kind of divination with scientific appeal. [230]

Sallustius, a Roman apologist for pagan beliefs and a friend of Emporer Justinian, during the early Christian era, commented on fate and evil in a manner that would be borrowed later by Christian apologists.

But to attribute men's acts of injustice and lust to Fate, is to make ourselves good and the Gods bad…..If Fortune makes a wicked man prosperous and a good man poor, there is no need to wonder. For the wicked regard wealth as everything, the good as nothing. And the good fortune of the bad cannot take away their badness, while virtue alone will be enough for the good.[231]

Sallustius used reason to argue that astrology was not true.

Also, the fact that the stars predict high or low rank for the father of the person whose horoscope is taken, teaches that they do not always

---

[230] Walter Burkert. *Greek Religion: Archaic and Classical* (trans. by John Raffan) Oxford: Basil Blackwell (1985), p. 329.

[231] From Sallustius. *On the Gods and the World*, ix "On Providence, Fate and Fortune" contained in Gilbert Murray. *Five Stages of Greek Religion* New York: Columbia University Press (1930), pp. 253-255.

make things happen but sometimes only indicate things. For how could things which preceded the birth depend upon the birth? [232]

## RELIGIOUS UNBELIEF

Almost from the beginnings of science with Thales, the existence of the gods of popular mythology were called into question. Initially, this disbelief was limited to philosophers who pursued the sciences or speculated on the nature of things. The general population was uneducated and still very superstitious, believing in divination through oracles, omens, dreams, and astrology. Ultimately, the disbelief became more widespread due to the availability of writings by philosophers the teachings of the sophists, and the plays of the poets. For example, Sextus Empiricus, a Greek skeptic, summarized the argument against the existence of the gods as follows:

> Anyone who says that there are gods says either that they provide for the things in the universe or that they do not – and that if they provide, then either for all things or for some. But if they provided for all things, there would be nothing bad or evil in the universe; but they say that everything is full of evil. Therefore the gods will not be said to provide for everything. But if they provide for some things, why do they provide for these and not for those? Either they both want to and can provide for all, or they want to but cannot, or they can but do not want to, or they neither want to nor can. If they both wanted to and could, then they would provide for all; but they do not provide for all, for the reason I have just given; therefore it is not the case that they both want to and can provide for all. If they want to but cannot, they are weaker than the cause in virtue of which they cannot provide for the things for which they do not provide; but it is contrary to the concept of god that a god should be weaker than anything. If they can provide for all but do not want to, they will be thought to be malign. If they neither want to nor can, they are both malign and weak – and only the impious would say this about the gods.

---

[232] Ibid., p. 254.

The gods, therefore do not provide for the things in the universe. But if they have providence for nothing and have no function and no effect, we will not be able to say how it is apprehended that there are gods, since it is neither apparent in itself nor apprehended by way of any effects. For this reason too, then, it is inapprehensible whether there are gods. From this we can deduce that those who firmly state that there are gods are no doubt bound to be impious: if they say that the gods provide for everything, they will say that they are a cause of evil; and if they say that they provide for some things or even for none at all, they will be bound to say either that the gods are malign or that they are weak – and anyone who says this is clearly impious.[233]

Stoicism is generally considered to have embodied several of the ideas that were to become the foundations of early Christianity. The founder of Stoicism, Zeno of Citium (334-262 BCE), was said by Sextus Empiricus, to have argued in favor of the existence of the gods as follows:

Zeno propounded this argument also: - "One may reasonably honour the Gods; but those who are nonexistent one may not reasonably honour; therefore Gods exist.[234]

This argument, relying on the belief of the masses of people, as evidence of the truth of the proposition, was, and still is, commonplace. Euripides captures this sentiment in *The Bacchae*.

---

[233] Julia Annas and Jonathan Barnes (trans.) *Sextus Empiricus: Outlines of Scepticism* Cambridge University Press (1994), III, 9-12 (pp. 145-146).

[234] R. G. Bury (trans.). *Sextus Empiricus* Vol. III London: William Heinemann Ltd. (1968), p. 73, Book I, 133. Sextus Empiricus, as a Cynic, didn't support arguments in favor of or opposed to the existence of the gods. See Edwyn Bevan. *Later Greek Religion* London: J. M. Dent & Sons Ltd. (1927), p. xxiii.

It is important to realize that Greek Sceptics raised no protest against the existing practice of religion. They did not contend that the arguments against the existence of gods were in the least stronger than the arguments in favour of popular beliefs. Their point was that the arguments for and against were precisely balanced. It was just as likely that there were gods as that there were not.

The simple nameless herd of humanity
Have deeds and faith that are truth enough for me.[235]

Sextus Empiricus refers to this approach as "arguing ... from the universal agreement of mankind."[236] Lucian (120-200 CE), in *Zeus Rants,* made fun of this logic in the dialogue between Timocles (a Stoic philosopher) and Damis (an Epicurean).

**Timocles.** See whether I frame this syllogism logically and whether you can capsize it in any way. If there are altars, there are also gods; but there are altars, ergo there are also gods. What have you to say to that?
**Damis**. After I have laughed to my heart's content I'll tell you.[237]

Sextus Empiricus summarized the confusion of religious beliefs – and disbelief – in ancient Greece in a manner that was critical of all of them.

Matters of piety and service to the gods are also full of much dispute. Most people say that there are gods, but some say that there are not – such as Diagoras of Melos, and Theodorus, and Critias of Athens. Of those who assert that there are gods, some believe in the traditional gods, others in those invented in the Dogmatist's schools – as Aristotle said that god is incorporeal and the limit of the heavens, the Stoics that he is breath pervading even loathsome things, Epicurus that he is anthropomorphic, Xenophanes that he is an unfeeling sphere. Some think that the gods provide for us, others that they do not. For what is blessed and indestructible, Epicurus says, takes no trouble itself and causes none for others. Again, among ordinary people, some say that there is one god, others that there are many and of different forms, so that they even fall in with the suppositions of the Egyptians who consider the gods to be dog-faced and hawk-shaped and cows and crocodiles and just about anything else.[238]

---

[235] Gilbert Murray. *Euripides and His Age* New York: Henry Holt and Co. (1913), p. 193.
[236] R. G. Bury (trans.). *Sextus Empiricus*, Op. Cit., p. 35, Book I, 60.
[237] A.M. Harmon. *Lucian* 8 vols. London: William Heinemann Ltd. (1953) Vol. II "Zeus Rants", p. 167.
[238] Julia Annas and Jonathan Barnes (trans.) *Sextus Empiricus: Outlines of Scepticism* Cambridge University Press (1994), III, 218-

Sextus Empiricus goes further and concludes that the lack of any agreement about the gods throws their existence in doubt.

> Now, since some of the Dogmatists say that god is a body, others that he is incorporeal, some that he is anthropomorphic, others not, some in space, others not – and of those who say that he is in space, some say that he is within the universe, others that he is outside of it – how shall we be able to acquire a conception of god if we possess neither an agreed substance for him nor a form nor a place in which he is? Even granting that god is indeed conceivable, it is necessary to suspend judgement about whether gods exist or not, so far as the Dogmatists are concerned. For it is not clear that gods exist: if the gods made an impression on us in themselves, the Dogmatists would be in agreement as to what they are and of what form and where; but the undecidable dispute has made it seem to us that the gods are unclear and in need of proof.[239]

To be sure, there were other factors, besides the birth of science, that caused skepticism about popular religion – especially among the educated. This is evident, for example, when the names of the important Greek astronomers are compared with the names of the most notable religious skeptics of the same period. These are, for the most part, entirely different groups.

Many of the challenges to religious beliefs based on ancient myths made by Greek philosophers and sophists were founded on moral and rational argument rather than science. The first sophist, Protagoras, said that the existence of the gods could not be proved by argument. This assertion provoked others that followed to claim that the gods were simply inventions of man. Critias of Athens wrote that the gods were created by clever people to keep order in civil society. Euhemerus wrote that

---

219 (pp.200-201).

[239] Ibid., III, 3,6 (p. 144).

gods were simply heroes and kings of ancient times who were worshipped for their famous deeds.

By the beginning of the fourth century BCE, disbelief in the gods of mythology was widespread among the educated. Evidence of this can be seen in a remark attributed to Antisthenes, the founder of the Cynic school of philosophy.

(Antisthenes) was born at Athens, but his mother was a Thracian woman. The fact that he was only half Greek is one of some importance in the history of Cynicism. It must, at any rate, have made it easier for him to break with accepted standards, religious as well as social. A full-blooded Hellene, even if he had shared Antisthenes' exclusive belief in a single supreme deity, would hardly have permitted himself the blasphemous exclamation, "If I could but lay my hands on Aphrodite, I would shoot her – that is, with the bow and arrows of her son.[240]

Strabo, the geographer, dismisses the gods of mythology as merely stories used by leaders to frighten the general population and keep order.

Now every illiterate and uneducated man is, in a sense, a child, and, like a child, he is fond of stories; and for that matter, so is the half-educated man, for his reasoning faculty has not been fully developed, and, besides, the mental habits of his childhood persist in him. Now, since the portentous is not only pleasing, but fear-inspiring as well, we can employ both kinds of myth for children, and for grown-up people too…Most of those who live in the cities are incited to emulation by the myths that are pleasing, when they hear the poets narrate mythical deeds of heroism, such as the labours of Heracles, or of Theseus, or hear of honours bestowed by gods, or, indeed, when they see paintings or primitive images or works of sculpture which suggest any similar happy issue of fortune in mythology; but they are deterred from evil courses when, either through descriptions or through typical representations of objects unseen, they learn of divine punishments, terrors, and threats – or even when they merely believe

---

[240] Theodor Gomperz. *Greek Thinkers: A History of Ancient Philosophy* 2 vols. (trans. by G. G. Berry) New York: Charles Scribner's Sons (1905) Vol. II, pp. 142-143.

that men have met with such experiences. For in dealing with a crowd of women, at least, or with any promiscuous mob, a philosopher cannot influence them by reason or exhort them to reverence, piety and faith; nay, there is need of religious fear also, and this cannot be aroused without myths and marvels. For thunderbolt, aegis, trident, torches, snakes, thyrsus-lances, - arms of the gods – are myths, and so is the entire ancient theology. But the founders of states gave their sanction to these things as bugbears wherewith to scare the simple-minded.[241]

## SCIENCE AND RELIGION

The conflict between science and religion in ancient Greece had its origins at Miletus around 600 BCE. At that time, the invention of science, with the discoveries of Thales and others, inevitably led to a revolution in the way Greeks regarded their world and, ultimately, undermined belief in the popular mythology about the gods. Knowledge of nature eventually displaced superstition. This was a gradual process, for the most part but these changes had a profound effect on society. Simply put, science changed the way man thought about everything for all time.

Philosophy is born when rational curiosity gets the upper hand, and rejecting the action of personal agents, men begin to seek the explanation of phenomena in the working-out of impersonal forces. The caprice of individual gods gives way to the search for general laws. In Greece, mythological and religious explanations were unchallenged until about 600 B.C., yet by the middle of the 4th century Aristotle, with Plato behind him, could move as easily as we do ourselves in the field of abstract conceptions and universal principles.
Before this could happen a great deal of hard thinking and reasoned argument had had to take place. But the point is this. There is no sudden transition from a mythical to a rational mentality. Mythical thinking does not die a sudden death, if indeed it ever dies at all. One has only to consider that Aristotle himself, the founder both of

---

[241] Strabo. *The Geography of Strabo* 8 vols. (trans. by Horace Leonard Jones) London: William Heinemann Ltd. (1960), Vol. I, 1.2.8 (pp. 69-71).

biological science and of formal logic, believed to the end of his days that the stars were living and divine.[242]

Ancient Greek astronomy was uniquely important in the history of the conflict between science and religion – not because astronomy involved the study of the heavens, where the visible gods were believed to reside, but simply because it was the first science.

The phenomena of the heavens are always adduced as a main source of belief in gods. But this belief is not of popular origin. Of the Greek natural sciences astronomy was that which had developed farthest and was studied most eagerly; this is the true reason why the heavenly bodies became involved in the discussion of religious and scientific questions and occupied a prominent place therein. Their cult only attained popular significance when astrology under Oriental influence pressed forward into a position of prominence; philosophy had prepared the way by impressing upon the mind of the public, and more particularly of the educated public, the idea of the special claim of the heavenly bodies to divinity. [243]

Observations of the heavenly bodies, begun by the Babylonians, led to the development of calendars which permitted the Greeks to plan cultivation of crops and navigate safely at sea. The fixed positions of the stars enabled man to determine his precise location on the globe and accurate distances to other cities. It was not long before certain Greeks concluded that the stars were only burning rocks and not immortal gods. This discovery, attributed to Anaxagoras, precipitated one of the first battles in the conflict between science and religion and Anaxagoras became one of the first casualties of the conflict.

---

[242] W.K.C. Guthrie. *In the Beginning: Some Greek Views on the Origins of Life and the Early State of Man* Ithaca, NY: Cornell University Press (1957), p. 15.
[243] Martin P. Nilsson. *A History of Greek Religion* (trans. by F. J. Fielden) 2nd ed., New York: W. W. Norton & Company, Inc. (1964), pp. 267.

The early scientists did not set out to challenge the myths about the gods but it was an inevitable consequence of the use of reason to understand the world. Indeed, most natural philosophers were pious believers in popular religious traditions. However, it was predictable that belief in supernatural forces was ultimately incompatible with the new knowledge provided by scientific discoveries. Also, in time, history replaced mythology and philosophy challenged religion.

The absence of an established priesthood, creeds, or sacred writings, in ancient Greece, helps explain why there was no strong public reaction to the religious implications of the first scientific discoveries.[244] Moreover, the fact that most scientific writings were only circulated among a small number of educated Greeks served to limit the immediate impact of these ideas on society. Later, scientific ideas were disseminated more widely by sophists who taught children from wealthy families and by poets who authored popular plays. In this way, science was introduced to the wider public and, by 400 BCE, was largely responsible for a significant movement away from traditional religion.

The natural philosophers, or first scientists, objected to supernatural explanations given for natural phenomena such as lightning or eclipses. The natural philosophers originated with the Milesian school – Thales,

---

[244] Sarah Iles Johnston (ed.). *Religions of the Ancient World: A Guide* Cambridge, MA: Harvard University Press (2004), p. 633.
That Greek religion lacked a canon, a clergy, and a spiritual center hardly comes as a surprise. Throughout its history, ancient Greece was a loose conglomerate of sovereign city-states, whose religious traditions and practices varied considerably from one place to another. Despite its Panhellenic tendencies, Greek religion retained a distinctly local and regional character.

Anaximander and Anaximenes. During the sixth century BCE, the writings of the natural philosophers did not lead to charges of impiety even though science left no role for the gods. However, by the end of the fifth century BCE, the public reacted quite differently. At that time, the Peloponnesian War was going poorly for Athens, there were threats to the democracy from the powerful, wealthy families, and there was a devastating plague. As a result, the public concluded that their gods were angry.

It was the discoveries of astronomy and other sciences that shifted the challenge to religion from moral to rational arguments. The weight of the evidence in nature argued against supernatural explanations.

> The consequences which the new cosmology brought about for religious thought were profound and extensive...In the new scheme of the world, which put the earth hanging free in space, there was not room for the underworld, the realm of the dead under the surface of the earth.[245]

Eventually, some philosophers became skeptical, not only about the nature of the gods, but also about their very existence. Cicero neatly drew a distinction between the popular superstition that the gods cause phenomena such as storms or floods and the actual objective existence of the gods.

> As to natural disturbances by land and sea, one cannot deny that when they occur they strike terror into many people, who attribute them to the power of the gods. But the question whether the gods exist or not is not the same as the question whether a number of people believe in their existence.[246]

---

[245] Gilbert Murray. *Four Stages of Greek Religion* New York: Columbia University Press (1912), p. 101.

[246] Cicero. *The Nature of the Gods* (trans. by Horace C. P. McGregor) New York: Penguin Books (1972), Book III, 16-17 [Balbus is speaking].

Plato squarely placed the blame for religious disbelief on science.

It is rather the novel views of our modern scientists that we must hold responsible as the cause of mischief. For the result of arguments of such people is this, that when you and I try to prove the existence of the gods by pointing to these very objects – sun, moon, stars, and earth – as instances of deity and divinity, people who have been converted by these scientists will assert that these things are simply earth and stone, incapable of paying any heed to human affairs, and that these beliefs of ours are speciously tricked out with arguments to make them plausible.[247]

Plato elaborated on this theme as described by historian Hugh Lloyd-Jones.

Unbelief, says Plato in the tenth book of the *Laws*, results from a combination of the scepticism engendered by such cosmological speculations as the materialistic theory of Anaxagoras about the heavenly bodies with the relativism in ethics promoted by such thinkers as Protagoras. The increasing familiarity of educated persons with views about the cosmos not at first sight easily reconcilable with traditional beliefs, or with the worship of heavenly bodies which Plato so warmly recommended, was no doubt a factor in the growth of scepticism.[248]

Later, Epicurus also expressed his disdain for science as a world without god.

For indeed, it were better to follow the myths about the gods than to become a slave to the destiny of the natural philosophers: for the former suggests a hope of placating the gods by worship, whereas the latter involves a necessity which knows no placation.[249]

---

[247] R. G. Bury (trans.). Vol XI "Laws" in *Plato* 12 vols. London: William Heinemann Ltd. (1968), Book X, 886D-E (p. 303). Spoken by Athenian stranger to Clinias of Crete. This discussion by Plato is evidence of widespread religious disbelief in his times.

[248] Hugh Lloyd-Jones. *The Justice of Zeus* Berkeley, CA: University of California Press (1971), pp. 129-130.

[249] Cyril Bailey (trans.). *Epicurus: The Extant Remains* Oxford: Clarendon Press (1926), p. 91.

The conflict between ancient Greek astronomy and pagan religion is one of perspective. Some Greek philosophers and writers considered the religious to be merely superstitious by denying natural causes for phenomena in the universe that are discoverable through the sciences. Conversely, other philosophers believed that the mere existence of the universe was proof positive of the existence of the gods and that the skeptics who disbelieve are incomprehensibly rejecting what is obvious. Both parties viewed the same evidence in nature and reached far different conclusions.

Curiously, it was not belief in myths of the gods but another sort of superstition that retarded scientific progress after its astonishing beginnings. The historian Cary sums up the impact of astrology quite succinctly.

The record of Hellenistic science shows a few notable gaps, such as the lack of systematic work in geology, which might have been expected to grow out of geography, and of the study of chemistry, whose fundamental problems had already been raised by the 'physicists' of the classical period. But the most remarkable feature in its history was its failure to lead on from its first successes to a continuous growth like that which modern pioneers of science inaugurated. This general falling off from earlier promise can hardly be attributed to ignorance of scientific method. The best specimens of Hellenistic research reveal precisely that blend of observation and reasoning which usually yields the most solid results in scientific discovery. A more valid if partial explanation is that Hellenistic science lacked instruments of precision, such as the telescope and microscope, and this at a stage when observation by the unaided eye could not carry them much further. This deficiency would help to explain the eventual stagnation of astronomy, botany and medicine. But the chief reason is, no doubt, to be found in the general loss of the spirit of hope and adventure which befell the Greek nation in the second century. It is significant that in the second century [BCE] the Greek world became addicted to the practices of astrology and magic. In pursuing these delusive short cuts to success it proved that it had lost the faculty of patient research which distinguished the earlier Hellenistic pioneers.[250]

The discoveries of astronomy and the other sciences were not, in themselves, the cause of the conflict between science and religion but rather it was the development of the scientific method with its emphasis on reason.

The real clash took place between that part of religion which interfered most in practical life and with which everyone came in contact every day, namely, the art of foretelling the future, and the attempts of natural philosophy to give physical explanations of celestial and atmospheric phenomena, of portents, and other events. Such explanations undermined the belief in the art of the seers and made it superfluous. For if these phenomena were to be explained in a natural way, the art of the seers came to naught. Belief in the oracles also was weakened.[251]

Observations and reason produced arguments that ultimately challenged the basic beliefs of the pagan religions. Moral arguments against pagan religious beliefs founded on the myths of the gods caused many to question the nature of the gods but it was rational argument which eventually led some to reject even the existence of the gods. The origins of both science and the conflict between science and religion are in ancient Greece. This conflict continues unabated today throughout the modern world.

---

[250] M. Cary. *A History of the Greek World: from 323 to 146 B.C.* Rev. 2nd ed. London: Methuen & Co. Ltd. (1951), pp. 352-353.
[251] Martin P. Nilsson. *Greek Folk Religion* New York: Harper & Brothers (1961), p. 136.

# CHAPTER 5
# CONCLUSION

Tantum religio potuit suadere malorum[252]
- Lucretius De Rerum Natura

Many modern historians trace the origins of the conflict between science and religion only to the time of Galileo and his trials by the church of Rome. This was the beginning of modern science which was made possible by the development of precision instruments for observation and measurement (e.g., the telescope). Certainly, the conflict between modern science and religion was most pronounced during that period. The establishment of hierarchical religious institutions with rigid doctrines based on sacred writings collided with the discoveries of the sciences about the nature of the universe.

These philosophies were directly opposed and ultimately incompatible. However, this conflict actually dates to a much earlier period. Reason alone suggests that the beginnings of the conflict between science and religion were coincidental with the invention of the first sciences c. 600 BCE in ancient Greece. There is also considerable evidence in the surviving writings of Greek poets, philosophers, sophists and historians that describe the beginnings of the conflict.

---

252 "So great are the evils to which religion has led men." Lucretius, *The Nature of Things*. Cited by Will Durant. *The Life of Greece* New York: Simon and Schuster (1966), p. 413. Lucretius was a Roman poet (c. 95 BCE-52 BCE) that expounded the philosophy of Epicurus.

Much has been written about the influence of politics, economics and philosophy on religion in ancient Greece and very little attention has been given to the topic of science and religion during this period. However, it is certain that the conflict between science and religion, beginning with its origins in ancient Greece, had a profound influence on thinkers of subsequent ages and even in modern times.

In the beginning, science was considered a subject for inquiry by philosophers along with medicine, logic, ethics, politics and much more. Thinkers who studied the nature of things in the physical world were known as natural philosophers. Astronomy was a very important early science because of its usefulness in developing a calendar for cultivation of crops or fixing a ship's position for seaborne trade. Some initial discoveries in astronomy were the source of the first conflict between science and religion – for example, the determination that the heavenly bodies were earth-like matter and not visible living gods.

To be sure, there were other sources of conflict with religion in ancient Greece but none had the influence that science did on the future development of society. Science irrevocably changed the way man thought about himself and his world. With each new discovery, the power of the gods was diminished and the stature of man was elevated. In a sense, man became god with the aid of science.

Even with the knowledge provided by science, the myths of the gods retained a strong influence on the minds of most men. There were attempts to rationalize these myths with the new knowledge derived from science. For instance, some asserted that the myths were allegorical or that they depicted a history of early heroes that was distorted by time. However, religion was merely weakened further by the efforts of apologists. In the cities, and especially among the educated, the influence of religious myths declined until new religions and philosophies replaced them.

Moral criticism of the gods of mythology, described by Homer and Hesiod, was another particularly powerful source of conflict with Greek religion. Initially, this criticism addressed anthropomorphism, idolatry, and blood sacrifices. In time, moral criticism would focus on issues such as the problem of evil and the capriciousness of the gods. This criticism, combined with the conflict between science and religion, eventually proved too much for religion based on the gods of mythology.

It is difficult to comprehend why outrageous human behavior was attributed to the ancient Greek gods of mythology unless it is understood, in large part, as an attempt to account for the existence of evil in the world. Absent these myths, the world simply presented a hostile environment, the past was inexplicable and the future a fearful mystery. The Pandora myth is another, more explicit, explanation for evil. Most often, evil is believed to be punishment of man by the gods for wrong conduct.

The preoccupation of Greek philosophies with the subject of right living is related to another aspect of the problem of evil. Simply, man is credited with free will and his wrong choices may lead to punishment by the gods. Impiety was a crime with consequences for both the individual and the society that failed to punish it since the gods were believed to protect the state and citizens from harm by external enemies.

The introduction of the concept of free will to answer questions about the existence of evil raised new questions about the efficacy of prayer and worship since foretelling by oracles implied predestination. Also, free will undermined the belief that the gods were omnipotent.

Traditional religion, even with the concept of free will, provided no comfort against the existence of evil. However, science yielded knowledge about the nature of things that explained mysteries formerly attributed to acts of the gods. The awareness of natural causes for phenomena such as weather replaced superstitious fears of the unknown. Predictably, for philosophers, knowledge and understanding became the goals of a good life.

The challenge to the gods of mythology from moral criticism and science was abundantly evident in writings from ancient Greece that have survived. That science carried the conflict is indicated by the gradual abandonment of the popular religion and the introduction of new religions and philosophies from Asia. The secular nature of western civilization at the present time is very much the product of the re-discovery of ancient Greek writings by the thinkers of Renaissance Europe.

Clearly, the world views provided by science and religion are directly opposed. Science embraces knowledge derived from observations of nature and the application of reason. Religion promotes an understanding based upon tradition and belief. Science is the study of the natural world and religion is the study of the supernatural.

In ancient Greece, after the discoveries of the first sciences, it was only possible to retain beliefs in the myths of the gods by suspending reason. But, even then, it was supposed that there were limits to man's knowledge that could only be explained by the existence and acts of the gods. So it is in the present day. Science does not yet appear capable of providing satisfactory answers to the ultimate questions regarding the origin of the universe and – more pointedly – the meaning of life. As a result, there continues to be a role for religion in society.

As previously mentioned, most historians describe the conflict between science and religion as of relatively recent origin – dating only to the beginning of the modern sciences in the Renaissance period. These historians often argue that there really was no conflict, except in a very limited sense, and even assert that religion promoted the development of science. However, these conclusions ignore the early history of the origins of the conflict between science and religion in ancient Greece. The absence of priests, doctrine and sacred writings in ancient Greece made the character of the early conflict different but there was a conflict of world views and considerable unrest in society nevertheless.

It should be demonstrated by this brief history of the origins of the conflict between science and religion in ancient Greece that many of the same issues confronting society then have persisted in modern times. The conflict itself may be viewed as a manifestation of the turmoil at the boundary between the known and unknown. This boundary has shifted but will always remain.

# Bibliography

Although it must be obvious to the reader by scanning the footnotes, I must here acknowledge my special indebtedness to the volumes by Thomas L. Heath, A. B. Drachmann and Martin P. Nilsson. Throughout, I have used English translations from the Loeb Classical Library, for the most part, since these are widely accessible.

Julia Annas and Jonathan Barnes (trans.). *Sextus Empiricus: Outlines of Scepticism* Cambridge: Cambridge University Press (1994)

Cyril Bailey (trans.). *Epicurus: the Extant Remains* Oxford: Clarendon Press (1926)

Edwyn Bevan. *Later Greek Religion* London: J. M. Dent & Sons Ltd. (1927)

John Burnet. *Early Greek Philosophy* 4th ed. Cleveland, OH: The World Publishing Co. (1964)

Paul Cartledge. *Democritus* New York: Routledge (1999)

Marcus Tullius Cicero. *The Nature of the Gods* (trans. by Horace C. P. McGregor) London: Penguin Books (1972)

C. Collard, M. J. Cropp and K. H. Lee. *Euripides: Selected Fragmentary Plays* Warminster, England: Aris & Phillips Ltd. (1995)

J. M. Cook. *The Greeks in Ionia and the East* New York: Frederick A. Praeger (1963)

Franz Cumont. *Astrology and Religion Among the Greeks and Romans* New York: G. P. Putnam's Sons (1912)

D. R. Dicks (ed.). *The Geographical Fragments of Hipparchus* London: University of London (1960)

Diels-Kranz (ed.). *The Older Sophists* (trans. by Rosamond Kent Sprague) Columbia, SC: University of South Carolina Press (1972)

A. B. Drachmann. *Atheism in Pagan Antiquity* Chicago, IL: Ares Publishers Inc. (1977) [reprint of 1922 ed.]

Adelaide Glynn Dunham. *The History of Miletus: Down to the Anabasis of Alexander* London: University of London Press (1915)

Will Durant. *The Story of Civilization: Part II The Life of Greece* New York: Simon and Shuster (1939)

Sextus Empiricus. *Against the Physicists* (trans. by R. G. Bury) Cambridge, MA: Harvard University Press (1968) [Loeb Classical Library]

William Fletcher (trans.). *The Works of Lactantius* Edinburgh: T&T Clark (1909)

P. M. Fraser. *Eratosthenes of Cyrene* London: Oxford University Press (1970)

Kathleen Freeman. *Ancilla to the Pre-Socratic Philosophers* Cambridge, MA: Harvard University Press (1977)

H. B. Gottschalk. *Heraclides of Pontus* Oxford: Clarendon Press (1980)

W.K.C. Guthrie. *A History of Greek Philosophy* Cambridge, England: Cambridge University Press (1962)

Moses Hadas (ed.). *The Complete Plays of Aristophanes* New York: Bantam Books (1962)

Thomas L. Heath. *Greek Astronomy* New York: Dover Publications Inc. (1991) [reprint of 1932 ed.]

T. L. Heath (ed.). *The Works of Archimedes* New York: Dover Publications Inc. [reprint of 1897 ed.]

Albert Henrichs. "Two Doxographical Notes: Democritus and Prodicus on Religion" *Harvard Studies in Classical Philology* Vol. 79, (1975)

Herodotus. *The History* (trans. by David Grene) Chicago, IL: The University of Chicago Press (1987)

Hippolytus. *Philosophumena or the Refutation of All Heresies*(trans. by F. Legge) London: Society for Promoting Christian Knowledge (1921)

Horace Leonard Jones (trans.). *The Geography of Strabo* London: William Heinemann Ltd. (1960) [Loeb Classical Library]

Charles H. Kahn. *Anaximander and the Origins of Greek Cosmology* New York: Columbia University Press (1960)

Charles H. Kahn. *The Art and Thought of Heraclitus* London: Cambridge University Press (1979)

Jan Fredrik Kindstrand. *Bion of Borysthenes: A Collection of the Fragments* Stockholm: Almquist & Wiksell Int'l. (1976)

Diogenes Laertius. *Lives of the Eminent Philosophers* (trans. by R. D. Hicks) Cambridge: Cambridge University Press (1965)

J. H. Lesher (trans.). *Xenophanes of Colophon: Fragments* Toronto: University of Toronto Press (1992)

Lucretius. *The Nature of Things* (trans. by Frank O. Copley) New York: W. W. Norton & Co. (1977)

E. C. Marchant (trans.) *Xenophon: Memorabilia and Oeconomicus* London: William Heinemann Ltd. (1938)

Richard McKeon (ed.). *The Basic Works of Aristotle* New York: Random House (1941)

Gilbert Murray. *Five Stages of Greek Religion* New York: Columbia University Press (1930)

O. Neugebauer. *A History of Ancient Mathematical Astronomy* New York: Springer-Verlag (1975)

Martin P. Nilsson. *A History of Greek Religion* (trans. by F. J. Fielden) 2nd ed. New York: W. W. Norton & Co (1964) [orig. published 1925]

C. H. Oldfather (trans.). *Diodorus of Sicily* London: William Heinemann Ltd. (1970) [Loeb Classical Library]

Bernadotte Perrin (trans.). *Plutarch's Lives* 11 vols. London: William Heinemann Ltd. (1932) [Loeb Classical Library]

Plato. *The Laws* (trans. By R. G. Bury) Cambridge, MA: Harvard University Press (1984) [Loeb Classical Library]

Plato. *Republic* (trans. By B. Jowett) New York: Modern Library (1982)

Plato. *Theaetetus* (trans. By Harold North Fowler) Cambridge, MA: Harvard University Press (1987)

Pliny. *Natural History* (trans. by H. Rackham) 10 vols. Cambridge, MA: Harvard University Press (reprinted 1967) [Loeb Classical Library]

Plutarch. *The Lives of the Noble Grecians and Romans* (Dryden translation rev. by Arthur Hugh Clough) New York: The Modern Library (1992)

Giorgio de Santillana. *The Origins of Scientific Thought 600 BC to 300 AD* Chicago, IL: University of Chicago Press (1961)

George Sarton. *A History of Science: Ancient Science Through the Golden Age of Greece* Cambridge, MA: Harvard University Press (1952)

Rosamond Kent Sprague (ed.). *The Older Sophists* Columbia, SC: University of South Carolina Press (1972)

W. W. Tarn. *Hellenistic Civilization* 2nd ed. London: Edward Arnold & Co. (1930)

Ivor Thomas (trans.) *Selections Illustrating the History of Greek Mathematics* 2 vols. Cambridge, MA: Harvard University Press (1993) [reprint of 1941 edition with revisions]

Thucydides. *History of the Peloponnesian War* (trans. By Charles Foster Smith) London: William Heinemann Ltd. (1952)

Albert Van Helden. *Measuring the Universe: Cosmic Dimensions from Aristarchus to Halley* Chicago, IL: The University of Chicago Press (1985)

Paul Veyne. *Did the Greeks Believe in their Myths?: An Essay on the Constitutive Imagination* (trans. By Paula Wissing) Chicago; IL: University of Chicago Press (1988)

Xenophon. *Memorabilia* (trans. By Amy L. Bonnette) Ithaca, NY: Cornell University Press (1994)

# EARLY GREEK ASTRONOMY

| GREEK ASTRONOMER | PRINCIPAL IDEAS/DISCOVERIES |
|---|---|
| Thales of Miletus<br>624-547 BCE | First astronomer. Predicted the solar eclipse of May 28, 585 BCE according to Herodotus. |
| Anaximander of Miletus<br>611-546 BCE | First natural philosopher. Speculated about the sizes of the sun and moon. Asserted that the Earth was unsupported by a medium since it is equidistant from everything in the universe. Taught that the universe was infinite and contained a plurality of worlds. Used a gnomon (sun dial) to predict the solstices, seasons, and equinox. First to draw a map of the inhabited part of Earth. |
| Anaximenes of Miletus<br>585-528 BCE | Concluded that stars give no heat because of their great distance from the Earth. Believed that there are other earth-like bodies [planets] in the universe. |

| GREEK ASTRONOMER | PRINCIPAL IDEAS/DISCOVERIES |
| --- | --- |
| Pythagoras of Samos<br>580-500 BCE | Discovered the sphericity of the Earth [also credited to Parmenides of Elia]. Determined that Venus is both the morning and evening star. |
| Anaxagoras of Clazomenae<br>500-428 BCE | Observed that the Moon was illuminated by the Sun which is a burning stone. Also, that the moon is earth-like with plains and mountains. Concluded that lunar eclipses are caused by the Earth's shadow. Trial and imprisonment for impiety and subsequently fined and banished from Athens for stating that the Sun was a burning stone. First victim of the conflict between science and religion. |
| Empedocles of Acragas<br>490-444 BCE | Concluded that light requires time to travel from the Sun to the Earth. |
| Philolaus of Croton<br>c. 470-390 BCE | Stated that the Earth, planets, Sun, and Moon revolved around a central fire [not the sun]. |

| GREEK ASTRONOMER | PRINCIPAL IDEAS/DISCOVERIES |
| --- | --- |
| Leucippus of Miletus<br>Fl. C. 450-420 BCE | First atomist. Taught that there were innumerable worlds created by the collision of small invisible particles of matter (atoms). |
| Democritus of Abdera<br>460-370 BCE | Atomist. Determined the order of the worlds: Moon, Sun, planets and fixed stars. Believed that the worlds differed in size and distance from each other. Asserted that some worlds have more than one sun and moon and some worlds are barren and without life. |
| Eudoxus of Cnidos<br>408-355 BCE | Developed a mathematical model of the universe consisting of rotating concentric spheres to explain the irregular motion of the planets. |
| Heraclides of Pontus<br>388-315 BCE | Taught that the Earth's rotation on its axis every 24 hours explains the apparent daily rotation of the heavenly bodies. Observed that Venus and Mercury orbit the Sun like satellites. |

| GREEK ASTRONOMER | PRINCIPAL IDEAS/DISCOVERIES |
|---|---|
| Aristarchus of Samos<br>310-230 BCE | First to propose the heliocentric hypothesis whereby the Earth orbits the Sun. Used observations to calculate the sizes and distances of the Sun and Moon from the Earth. |
| Archimedes of Syracuse<br>287-212 BCE | Famous mathematician and inventor. Calculated diameter of the Sun. Determined accurate value for pi. |
| Eratosthenes of Cyrene<br>276-195 BCE | Calculated the circumference of the Earth. Also calculated the angle of inclination of the Earth's axis. Introduced the Babylonian sexagesimal system of mathematical notation which led to angular measurements using degrees. |
| Hipparchus of Nicea<br>190-120 BCE | Father of astronomy. Measured lunar cycle and developed an accurate calendar. Tabulated first star catalogue with 850 stars (position and degree of brightness). |

| GREEK ASTRONOMER | PRINCIPAL IDEAS/DISCOVERIES |
|---|---|
| Poseidonius of Apamea<br>130-50 BCE | Calculated the circumference of the Earth and the sizes and distances of the Sun and Moon. |

<u>NOTE</u>: This is not a complete list of early Greek astronomers. Only the astronomers who made the most notable discoveries are shown.

# EARLY GREEK UNBELIEF

| GREEK THINKER | PRINCIPAL IDEAS |
|---|---|
| Xenophanes of Colophon<br>570-478 BCE | Believed that the myths about the gods in Hesiod and Homer were untrue. Wrote that the anthropomorphic gods were clearly and invention of man. Questioned whether it was possible for man to ever know anything about the gods. |
| Anaxagoras of Clazomenae<br>500-428 BCE | Did not deny the existence of the gods but attributed natural causes to astronomical phenomena. Imprisoned by Athenians for teaching that the Sun was a burning stone and the Moon was earth-like – not visible gods. Later, exiled and lived at Lampsacos in Asia Minor. Trial on charge of impiety widely considered a pretext for prosecuting him because Anaxagoras was a teacher and friend of Pericles who had many powerful political enemies in Athens. |
| Hippo of Rhegium<br>c. 5$^{th}$ century BCE` | Asserted that nothing existed – including the gods – unless it could be perceived by the senses. |

| GREEK THINKER | PRINCIPAL IDEAS |
| --- | --- |
| Heraclitus of Ephesus<br>c. 540-480 BCE | Criticized blood sacrifice as an offense against the gods and opposed prayers to statues as mere idolatry. Like Xenophanes, he found these to be inventions of man. |
| Diogenes of Sinope<br>c. 404-323 BCE | Early Greek naturalist and Cynic. Ascribed natural causes to phenomena and denied supernatural. Did not believe gods existed because evil went unpunished. Accused of atheism. |
| Diagoras of Melos<br>fl. c. 464 BCE | Decree issued at Athens in 415 BCE offered a reward for killing Diagoras for impiety because he divulged the Eleusinian mysteries. Believed that the gods were indifferent to human affairs because they failed to intercede to help man. |
| Protagoras of Abdera<br>fl. c. 450-430 BCE | The first sophist. Tried for impiety in Athens and exiled. His writings were burned. Asserted that he didn't know whether the gods existed. |

| GREEK THINKER | PRINCIPAL IDEAS |
| --- | --- |
| Democritus of Abdera<br>c. 460-370 BCE | Atomist who taught that there was a mechanical world comprised of very small, invisible atoms. Believed that all phenomena have natural causes that previously men had attributed to gods. No role left for gods. |
| Critias of Athens<br>480-404 BCE | Student of Socrates. Member of the Thirty Tyrants that overthrew the democracy and ruled Athens briefly. A character in his play *Sisyphus* stated that men invented the gods to prevent lawlessness. |
| Prodicus of Ceos<br>fl. c. 470 BCE | Stated that early man deified useful things and forces, e.g., Sun, water fire. |
| Bion of Borysthenes<br>fl. c. 300 BCE | Rejected religious belief and worship, including prayers and divination, as superstition. |
| Euripides of Salamis<br>480-406 BCE | In several of his plays, the characters question whether there are gods because evil prevails over good and is unpunished. Friend of Pericles and Protagoras. |

| GREEK THINKER | PRINCIPAL IDEAS |
| --- | --- |
| Socrates of Athens<br>468-399 BCE | Accused of being a sophist, like Protagoras, who were generally considered to be atheists because they questioned everything including the existence of the gods. Stood trial for impiety at Athens and forced to take poison in prison. |
| Aristotle of Stagira<br>384-322 BCE | Forced into exile to avoid prosecution at Athens for impiety because he wrote that Olympian gods were myth and not real. Considered a victim of anti-Macedonian sentiments in Athens after the death of Alexander. |
| Euhemerus of Messene<br>fl. c. 300 BCE | Wrote that the gods were originally men of great accomplishment who were later worshipped as gods. |
| Epicurus of Samos<br>341-271 BCE | Taught the naturalism of early atomists. Believed that the gods were remote and not interested in man. Considered by many to be an atheist. |

# INDEX


Abraham 2 (n.3), 23 (n.46)
Alexander the Great 56-57, 86
*Almagest* 42
Anaxagoras of Clazomenae
- impiety 21-22, 48-50
- lunar eclipse 21
- moon's surface earthlike 21
- Pericles 22, 50
- plurality of worlds 21
- sun a burning stone 20, 50 (n.100)

Anaximander of Miletus
- earth cylindrical 16
- first map 16-17
- gnomon 17
- infinite universe 17
- plurality of worlds 17

Anaximenes of Miletus
- distance of stars 18
- gnomon 18

Antisthenes of Athens
- unbelief in gods 124

anthropomorphism
- gods invented 47, 86
- myths impious 46
- Olympian gods 43-45, 110

Archimedes of Syracuse
- diameter of sun 31
- pi 31
- *The Sand-Reckoner* 27-28, 31 (n.63)

Aristarchus of Samos
- heliocentric hypothesis 3 (n.5), 27-29
- sizes and distances of moon, sun 30

Aristophanes of Athens
- *Birds* 57
- *The Clouds* 55, 58

Aristotle of Stagira
- gods are myths 85
- gods in form of men 85
- impiety 85-86
- sphericity of earth 19

astrology 3, 131
astronomy 1-4
atheists
- Bion of Borysthenes 67
- Critias of Athens 63
- Diagoras of Melos 57
- Diogenes of Sinope 56
- Epicurus of Samos 91 (n.181)
- Euhemerus of Messene 88
- Euripides of Salamis 69
- Hippo of Rhegium 53
- Protagoras of Abdera 59
- Socrates of Athens 79-84
- Stilpo of Megara 94, 106
- Strato of Lampsacos 94
- Theodorus of Cyrene 106

atomist cosmogony
- Leucippus of Miletus 24
- origins of theory 24 (n.51)

Babylonia 2, 10-11, 13
*Bacchae, The* 122
*Bellerphon* 69-70
Bion of Borysthenes
- rejected popular religion 67
- on punishing children of wicked 68

*Birds* 57
Blake, William
- *The Tiger* 112 (n.216)

book burning 60 (n.121), 105

catalogue of stars 39
Cicero (Marcus Tullius) 93, 113, 129
existence of gods 93, 129-130
gods and evil 56
irreligion 65-66
*City of God* 117
Clement of Alexandria 53
*Clouds, The* 55, 58, 79-80
colonies 5-7, 100
Columbus, Christopher 40
concept of causation 25
conflict between science and religion 52, 125-132
Copernicus, Nicholas 3 (n.5), 29 (n.59)
Critias of Athens
atheist 63
Sisyphus 63-64
Cynicism 67, 105
*Daughters of Troy* 71
Darwin, Charles 113 (n.218)
*De Natura Deorum* 93
*De Rerum Natura* 89
degrees 37
Delphi, oracle at 97-98 (n.190)
Democritus of Abdera
mechanical world 61
order of heavenly bodies 25
plurality of worlds 25
Diagoras of Melos
*Birds* 57
decree for impiety 57 (n.116)
"the Melian" 58
Diogenes of Sinope
Alexander the Great 56-57
on unpunished evil 56
*The Clouds* 55
Earth
center of universe 27
circumference 32-34, 39
moves in a circle 24
orbits the sun 27
revolves on its axis 27
size of land masses 32 (n.66), 35
sphericity 19, 34
eclipse
lunar 18, 21
Saros period 13
solar 38 (n.79), 48 (n.98)
ecliptic
defined 37 (n.78)
obliquity 37 (n.78)
Egypt 10
*Electra* 73
Empedocles of Acragas
four elements 23
travel time of light 23-24
Epicurus of Samos
*De Rerum Natura* 89
Multitudes beliefs about gods wrong 93
nature of gods 90-91
origins of evil 90-91
science and gods 130
equinox 37 (n.78)
Eratosthenes of Cyrene
circumference of earth 32-34
oceans connected 32
sexagesimal system 36
size of inhabited world 35
Eudoxus of Cnidos
irregular motion of planets 26
Euhemerus of Messene
gods as historical figures 86-87
heavenly bodies gods 88
*Sacred History* 86-87

Euripides of Salamis
atheist 69
death as refuge from evil 73
evil triumphs over good 71
evil unpunished 70, 73
existence of gods 74
god is nature or invention 71
gods do evil deeds 73, 109 (n.211)
prosecution for impiety 75-76
evil
cause of evil 112-113
death as refuge from evil 73
evil prospers 114-115
gods watch men 117
good and evil are one 55
problem of evil 111-118
punishment of evil 56, 70, 73, 114-118
Tityus 56
fate and religion 118-120
flood 44 (n.89)
geocentric hypothesis
*Almagest* by Ptolemy 42
gnomon
circumference of earth 32
latitude 37
maps 37
Greek religion
city life 99
fate 118-120
history 101-103
literature 107-108
moral criticism 108-111
oracle at Delphi 97-98
philosophy 103-105
problem of evil 111-118
science 125-132
sophists 105-107
state 96-98
temples 96
trade 100-101
unbelief 120-125
heavenly bodies
gods 23 (n.46), 49
irregular motion 26
order of 25
heliocentric hypothesis
Aristarchus 27
Copernicus 3 (n.5)
Plutarch 29 (n.59)
Seleucus the Chaldean 28 (n.58)
*The Sand Reckoner* 27-28
Heraclides of Pontus
earth rotates on axis 27
Mercury and Venus orbit sun 27
Heraclitus of Ephesus
good and evil 55
on idolatry 54
*river* metaphor 54-55
Herodotus
eclipse predicted by Thales 12
religion 101
Hesiod
*Theogony* 43
*Works and Days* 1
Hipparchus of Nicea
maps using degrees 37
mathematical geography 37
Hippo of Rhegium
atheism 53
*History of Peloponnesian War* 102
Homer
anthropomorphism 9
*Iliad* and *Odyssey* 45
Tityus 56

Hume, David 51 (n.103)
idolatry 54
impiety
Anaxagoras of Clazomenae 50-51, 97
Aristarchus of Samos 29
Aristotle of Stagira 85-86
Bion of Borysthenes 67
Critias of Athens 63
Democritus of Abdera 62
Diagoras of Melos 57
Euripides of Salamis 69
Plato of Athens 51
Prodicus of Ceos 67
Protagoras of Abdera 59
Socrates of Athens 81-84
*Inferno* 91 (n.181)
*Iphigenia in Tauris* 72
Josephus
*Jewish Antiquities* 2(n.3)
Lactantius 91 (n.181)
latitude
fixed by sun dial 35
used to draw maps 37
Leucippus of Miletus
atomist cosmogony 24
concept of causation 25
light, speed of 23-24
lightning 62
longitude
fixing by lunar eclipse 38 (n.79)
Lucian
*Zeus Rants* 74, 92
Lucretius 89, 133 (n.250)
lunar eclipse
Anaxagoras of Clazomenae 21
map of earth 16-17, 37
Mercury 27
Miletus
Ionian colony 5
origin of conflict between science and religion 125
trading center 5-8
moon
distance from earth 20, 38, 40
lunar orbit 40-41
size 30, 38, 40
surface earth-like 21
moral criticism of religion 108-111
obliquity of ecliptic 37
oceans
size 35-36
*On the Face in the Round of the Moon* 29 (n.59)
oracles 97-98 (n.190)
order of planets 25
Pandora 44
Pericles
defense of Anaxagoras 22-23 (n.45)
solar eclipse 48 (n.98)
Philodemus 66
Philolaus of Croton
central fire 24
pi
Archimedes 31
defined 31 (n.64)
Plato of Athens
science and religious unbelief 129
kinds of impiety 51 (n.103)
Pliny the Elder (Gaius Plinius Secundus)
Aegospotami meteorite 20 (n.42), 22 (n.44)
on gods 112
plurality of worlds
Anaximader of Miletus 17

Anaximenes of Miletus 18
Plutrch
Anaxagoras 14 (n.45), 21 (n.44), 122 (n.45)
*On the Face in the Round of the Moon* 29 (n.59)
squaring the circle 21 (n.44)
Poseidonius of Apamea
circumference of earth 39-40
distance of moon 40
phases of moon and tides 40
size of moon 40
priests 72, 127 (n.242)
Prodicus of Ceos
charged with impiety 65
man deifies things 65
Prometheus 44
Protagoras of Abdera
existence of gods 60 (n.121)
man is measure 59
Ptolemy 42
punishment after death
Tityus 56
Bion of Borysthenes 67
Pythagoras of Samos
central fire 19
distance to moon and sun 20
earth not at center of universe 19
sphericity of earth 19
Venus in morning and evening star 20
religious unbelief 120
*Republic* 80
revolution of planets 27
*river* metaphor 54-55
rotation of earth 27
Rule of Three 15 (n.28)
*Sacred History* 86-87
St. Augustine 117
Sallustius
astrology not true 120
on fate 119
*Sand-Reckoner, The* 27-28, 31 (n.63)
science
Epicurus on science 130
origins at Miletus 5-11
science and religion 125-132
sources of religious unbelief 129-130
scientific method 84
Seleucus the Chaldean 28 (n.58)
Semonides of Samos 107-108
Sextus Empiricus
confusion of religious beliefs 122-123
nonexistence of gods 120-121, 123
punishment of dead 115
sexagesimal system 36
Shelley, Percy Bysshe 95
*Sisyphus* 63-64
Socrates of Athens
gods not source of evil 80-81
on astronomy 106
portrayed by Aristophanes 79-80
trial for impiety 81-84, 106
sophists
Protagoras of Abdera 59-61
Prodicus of Ceos 65-67
sphericity of earth 19
squaring the circle
Anaxagoras of Clazomenae 22
pi 31

stade 33 (n.70), 36 (n.75), 37
star catalog 39 (n.81)
state and religion 96-98
Stilpo of Megara 94
stoics 41
Strabo of Amasia 7, 24 (n.51), 33 (n.69), 35-36
    gods are myths 124-125
Strato of Lampsacos 94
sun
    burning stone 20 (n.42), 41-42, 49
    distance 20, 30, 38-39
    fire 41-42
    size 14, 30
    solar parallax 39
Syene 33
Tennyson, Alfred Lord
    *In memoriam* 112 (n.216)
Thales of Miletus
    earth spherical 14
    height of pyramids 14-15
    size of sun's orbit 14
    solar eclipse 12-13
Theodorus of Cyrene 67
Theognis of Megara 114
*Theogony* 43
Thucydides
    destruction of Melos 102
    funeral speech of Pericles 102-103
    Peloponnesian War 101-103
    plague in Athens 103
tides
    oceans connected 32
    phases of moon 40
trade and religion 100-101
*Trojan Women* 71
Venus
    morning and evening star 20, 27
*Works and Days* 1-2
worlds, other 21, 25
Xenophanes of Colophon
    anthropomorphic gods 47
    knowledge 47
    myths of gods impious 46
Xenophon
    *Memorabilia* 69
Zeno of Citium 41-42
Zeus 43-44, 52, 71, 79, 86-87, 92
*Zeus Rants* 74, 92, 122

# About the Author

Craig Leisy is the author of other self-published volumes including, *The Blasphemy Trial of Charles B. Reynolds at Morristown, New Jersey on May 19-20, 1887*. In that book, and this one, the historical characters tell the story in their own words. Craig is preparing a volume of poetry for publication under the title, *Rain's Only Wet if You're In It*. Craig is also currently at work on another writing project that will examine the evidence of Epicureanism in the life and writings of Omar Khayyam, a medieval Persian mathematician, astronomer and poet, who gained lasting international fame with the publication of Edward FitzGerald's translations of the *Rubaiyat of Omar Khayyam* beginning in 1859.

Made in United States
Orlando, FL
04 November 2026

90201199R10096